J626

D1339669

Caloric Restriction: A Key to Understanding and Modulating Aging

RESEARCH PROFILES IN AGING

VOLUME 1

Caloric Restriction: A Key to Understanding and Modulating Aging

Edward J. Masoro
Charleston, South Carolina
USA

2002

ELSEVIER

Amsterdam – Boston – London – New York – Oxford – Paris
San Diego – San Francisco – Singapore – Sydney – Tokyo

ELSEVIER SCIENCE B.V.
Sara Burgerhartstraat 25
P.O. Box 211, 1000 AE Amsterdam, The Netherlands

First edition 2002

Library of Congress Cataloging in Publication Data
A catalog record from the Library of Congress has been applied for

British Library Cataloguing in Publication Data
A catalogue record from the British Library has been applied for

ISBN: 0-444-51162-8
ISSN: 1567-7184 (Series)

♾ The paper used in this publication meets the requirements of ANSI/NISO Z39.48-1992 (Permanence of Paper).
Printed at Netherlands.

TABLE OF CONTENTS

PREFACE

For many years, it has been known that mice and rats maintained on a long-term reduced food intake have an increased life span and remain healthy and vigorous at their advanced ages. What is the reason for this change in the usual pattern of aging? The evidence is overwhelming that the life extension results from a slowing of the aging processes, and the factor responsible is the decrease in caloric intake. The obvious question: How does this factor work? A good question—and the reason research on the anti-aging action of caloric restriction is today one of the most studied research areas in biological gerontology.

In 1988, Rick Weindruch and Roy Walford co-authored *The Retardation of Aging and Disease by Dietary Restriction*, which provided an excellent encyclopedic coverage of research findings, and played an important role in attracting talented investigators into the field. Indeed, in the ensuing years, research activity increased to such an extent that encyclopedic coverage would not only be a monumental undertaking but also one that would prove far too tedious to read. Yet, as new investigators enter the field and gerontologists and geriatricians seek to keep abreast of developments, what is needed is an update of the field that also provides its historical underpinnings. The present book aims to meet these needs.

A general history of the field is presented in the first chapter. In all of the other chapters, each of the many specific areas of the field is covered, starting with the earliest findings and focusing on the recent studies that provide significant new information. Of course, if findings are discordant, the differences are discussed. I believe this approach has permitted the coverage of the salient findings in all subject areas with a text that the reader will find succinct as well as useful and interesting.

It is my hope that this book will be a valuable resource for all those studying the biology of aging, including senior investigators, postdoctoral trainees, and predoctoral students, as well as academic geriatricians.

E. J. Masoro
Nantucket, MA, USA
August 18, 2002

CHAPTER 1

Overview

Contents

Extending length of life and slowing or even reversing aging have been a human preoccupation throughout recorded history (Birren, 1996). Undoubtedly, it was a human desire in preliterate times as well.

Alexander the Great (356–323 B.C.) described encountering a fountain of life, and Ponce de Leon (1460–1521) purportedly discovered Florida while seeking the fountain of youth, waters said by local Indian lore to have the power to rejuvenate. Over the years, a multitude of different substances have been claimed to prevent aging and extend life. In addition to such "magic bullets," manipulating lifestyle has been viewed as a means of slowing aging and increasing longevity. As early as the 14th century, Cornaro, an Italian author whose book *Discourses* greatly influenced thought in Western Europe, was way ahead of his time when he proposed that good hygienic practices would have such an effect.

Indeed, an astounding increase in human life expectancy in the "developed" nations occurred during the 20th century because of improved hygiene as well as other environmental and lifestyle factors. However, it must be pointed out that most, if not all, of this increase was not due to retarding aging. Rather, advances in nutrition, medicine, living standards, and public health enabled the population to age by preventing early death.

RESEARCH PROFILES IN AGING
VOLUME 1 ISSN 1567-7184

The extent to which such changes in environment and lifestyle have also delayed aging still remains to be defined. Indeed, in mammalian species, it has been remarkably difficult to clearly establish the validity of the many claims about manipulations that extend life by retarding aging.

A notable exception is restriction of caloric intake in rodent species. This experimental intervention, referred to as caloric restriction (CR) or dietary restriction (DR) or food restriction (FR), markedly and reproducibly extends the life span of rats and mice by slowing the aging processes. CR is the acronym that will be used in this book.

The anti-aging action of CR has become a major subject of research in biological gerontology. This is hardly surprising, given the long history of human interest in life extension and preservation of youth. However, most probably the major reason is CR's usefulness as a tool for the study of aging since it is a manipulation that markedly retards the aging processes in a number of species.

Historical background of the CR paradigm

Early in the 20th century, Osborne and Mendel (1915) and Osborne et al. (1917) reported that retarding the growth of rats by nutritional means resulted in an increase in length of life. Unfortunately, the longevity component of these studies was flawed by the premature loss of animals due to disease. Moreover, five years later, Robertson and Ray (1920) published a paper that showed the rate of growth of mice positively correlates with length of life.

In the course of their research on retardation of growth in trout due to a low protein diet, McCay et al. (1929) noticed that the fish with retarded growth outlived those that grew normally. This observation led McCay and associates to tackle the issue of lack of agreement in the rodent literature. In 1930, they started experiments with white rats to determine the effect of retarded growth on life span. In the design of their first study, McCay et al. (1935) allowed one group of rats to grow to maturity at a normal rate and slowed the growth of two other groups by restricting food intake. The restricted groups did not grow at all and, indeed, began to fail, whereupon the amount of food was increased just enough to keep the animals alive. Thus the restricted groups underwent long periods of no growth interspersed with spurts of growth. One of the two groups was maintained on the restricted diet regimen for 700 days and the other for 900 days, after which all were allowed to eat *ad libitum*. While the mean length of life of the rats that grew normally was about 600 days, many of the rats on the growth-retarding dietary regimes lived much longer. This first study had involved the reduced intake of all nutrients. In a second study, McCay et al. (1939) decreased the intake of calories but not of protein, minerals, or vitamins; the results of the second study were similar to those of the first study. Since the focus of McCay and his colleagues was on the influence of growth on longevity, it is not surprising that they concluded that the increase in length of life was due to the decrease in rate of growth.

From the 1940s through the 1960s, research on CR focused on the age-associated diseases of rats and mice. Saxton and Kimball (1941) reported that CR retarded the progression of chronic nephropathy in rats, a major age-associated disease in this species. And Saxton et al. (1944) found that CR inhibited the development of leukemia in mice. Ross (1961) and Ross and Bras (1965) carried out extensive research on the influence of CR on longevity and age-associated diseases of rats, with an emphasis on neoplasia and the possible role of specific nutrients. They concluded that the reduction in caloric intake increased longevity and retarded disease. When Berg and Simms (1960) undertook similar studies with rats, they proposed that reduction in body fat is responsible for the life extension and inhibition of age-associated diseases seen with CR.

Animal taxa studied

During the past 70 years, rats and mice by far have been the species most used in CR research. CR has been found to increase longevity of both genders of many different stocks and strains of mice and rats. [However, in one study it was reported that CR decreases the longevity of male C57BL/6J mice (Harrison & Archer, 1987).] In addition to its marked life-extending effect in rodent species, CR has also been found to increase the length of life of invertebrates—protozoa, flies, water fleas, nematodes, rotifers, and spiders (Austad, 1989; Fanestil & Barrows, 1965; Hekimi et al., 2001; Kirk, 2001; Klass, 1977; Lynch & Ennis, 1983; Rudzinska, 1951; Sawada and Carlson, 1987; Vanfleteren & Braeckman, 1999; Weindruch & Walford, 1988). It also extends the life span of other vertebrates, such as hamsters (Stuchlikova et al., 1975) and fish (Comfort, 1963; Weindruch & Walford, 1988). Does CR extend the life of mammals other than rats, mice, and hamsters? The question remains unanswered because well-designed research addressing the effect of CR on both mean and maximum length of life has not yet been published. Indeed, most mammalian species have such long lives that research on the influence of CR on their longevity is a long and expensive endeavor. In fact, for some species, the length of such a study would be longer than the scientific life span of an investigator. Thus, few investigators have been willing to undertake such studies.

However, Kealy et al. (2002) have recently reported that CR increases the mean length of life of dogs (Labrador Retrievers). Unfortunately, this study does not have the statistical power to determine the effect on maximal length of life.

Fortunately, there are ongoing long-term studies on the effects of CR on nonhuman primates. Rhesus monkeys are being studied in three laboratories (Ingram et al., 1990; Kemnitz et al., 1993; Hansen & Bodkin, 1993). A similar long-term study is also ongoing on squirrel monkeys (Ingram et al., 1990). Also, there is a short-term study of cynomolgus monkeys (Cefalu et al., 1997). Thus far, none of these nonhuman primate studies has been carried on long enough to yield information on the effect of CR on longevity of these monkeys (Roth et al., 1999). Hopefully, one or more of the studies will ultimately generate this information.

Meanwhile, the effects of CR have also been studied in humans, but such studies have usually been of short duration. About a 20% reduction of caloric intake for 10 weeks was found to decrease blood pressure and body fat mass and to increase HDL-cholesterol levels in human subjects (Velthuis et al., 1994). Biosphere 2 led to an interesting study by Roy Walford, one of the eight men and women who spent two years in the Biosphere. During most of the two years, the participants ate a CR diet dictated by necessity rather than design. Although less than a well-designed study (it was not designed as a CR study), the findings show that many physiological and biochemical effects of CR observed in rats and mice also occurred in these human subjects (Verdery & Walford, 1998; Walford et al., 1992, 2002).

Study designs

The design of the pioneering research of McCay and associates had been aimed at determining the effect of growth on longevity, and reducing caloric intake was merely the tool used to retard the growth of the rats. For this reason, they employed the complicated procedure described above, which reduced the caloric intake and thus the growth of a rat as much as possible without killing the animal. Since then, most research has focused on the influence of caloric intake on longevity, and several relatively simple study designs for reducing caloric intake have been employed in the rodent and the nonhuman primate studies.

Probably the most straightforward design consists of a control group that is fed *ad libitum* and one or more CR group(s) fed a reduced amount of food. In most studies, the CR groups were fed 30–50% fewer calories than the control group; however, a recent report indicates that a reduction of only 10% of caloric intake can significantly increase longevity in rats (Duffy et al., 2001). Procedures for achieving a 40% reduction in caloric intake in rats are presented in detail in a paper by Yu et al. (1985). Weindruch et al. (1986) employed a modified version of this design in a mouse study; the control group was fed 85 kcal/week (less than they would eat if fed *ad libitum*) and the CR group was fed 40 kcal/week. Merry & Holehan (1985) used body weight as a measure of caloric restriction in rats; the CR group was allotted the amount of food needed to maintain a body weight of 50% that of the *ad libitum* group. Still another method to reduce caloric intake by rats and mice is feeding every other day (Goodrick et al., 1982); Duffy et al. (1994) modified this procedure to achieve precisely a 40% reduction in caloric intake by mice. While many different methods have been used to reduce caloric intake in rats and mice, all have yielded similar findings. Although in most studies, CR was initiated early in life (1–3 months of age), starting CR at ages older than that has also been found to increase longevity (Weindruch, 1996).

Investigators studying CR in nonhuman primates have adopted modifications of the procedures used in rodent research. Ingram et al. (1990) started their study of rhesus monkeys with animals in two different age ranges (a 1-year-old group and a group of 3–5-year old). Their squirrel monkey study also involved two age groups (1–4-year old and 5–10-year old). In both studies, each age group was divided into control and CR subgroups, and over a period of three months the food intake of the CR subgroups was gradually reduced by 30%. Kemnitz et al. (1993) measured the food intake of rhesus monkeys 8–14-year old, and generated a CR subgroup by gradually reducing the food intake of each monkey in that subgroup to 70% of its baseline intake. Hansen and Bodkin (1993) started with rhesus monkeys 11–12-year old and weighing 10–12 kg. The control group was allowed to gain weight as they aged, while the CR group was maintained at 10–12 kg body weight by restricting food. The latter design is similar to that of McCay and colleagues, but is less difficult to execute because the monkeys were mature while McCay and associates started their rat study with weanlings, a rapidly growing stage of life.

Anti-aging action

CR is commonly said to slow the aging processes, i.e., to slow those deteriorative changes that result in an increasing vulnerability to challenges, thereby decreasing the ability of an organism to survive as adult chronological age increases. However, the fact that CR increases the mean length of life of a population does not necessarily mean it has slowed the aging processes. For example, although human population longevity increased markedly in the developed nations during the 20th century, much of that increase related to prevention of premature death of infants, children, and adults. Whether slowing of the aging processes has played any role in the increase in human longevity remains an open question (Robine et al., 1997). However, detailed analyses of mortality characteristics of mouse and rat populations in CR studies indicate that the slowing of the aging processes plays an important, if not the sole, role in CR's effect on longevity. These analyses are the subject of Chapter 2.

In addition, CR retards the age-associated deterioration of many physiological processes, and this is also a strong indication of an anti-aging action (Weindruch & Walford, 1988). The many effects of CR on cellular and molecular biology and systemic physiology are presented in Chapters 3 and 4, respectively.

Finally, CR delays the occurrence and/or slows the progression in severity of most of the age-associated diseases of rats and mice, including those unique to specific genotypes (Masoro, 1993). Since age-associated diseases must be considered an integral aspect of aging, this action also indicates that CR has an anti-aging action (Masoro, 1999; Semsei, 2000). The effects of CR on age-associated diseases are presented in Chapter 5.

Dietary factor

As mentioned above, the results of their second rat study led McCay and associates to conclude that the restriction of calories is the dietary factor responsible for the life-extending action of decreased food intake. Further support for this conclusion was provided by a subsequent study (McCay, 1942). A large group of rats was maintained on a low intake of a nutritionally adequate basal diet. However, four subgroups received additional calories: the source for one subgroup was dried liver; for another subgroup, whole milk; for a third subgroup, dried cooked starch; and for a fourth subgroup, sucrose. Each of the subgroups exhibited a similar decrease in length of life as compared to rats maintained on the low caloric intake of the basal diet throughout life.

The conclusion of McCay and associates was strengthened by more recent studies in which a range of semisynthetic diets was used, i.e., diets comprised of purified components, such as casein as the protein source;

corn oil as the fat source; dextrin (a readily digestible preparation from starch) as the carbohydrate source; and synthetic mineral mixes and vitamin mixes (Masoro, 1990). Such diets permit easy manipulation of individual dietary components. Indeed, CR studies have been conducted, in which, one single component of the semisynthetic diet was not restricted as follows: protein (Masoro et al., 1989); fat (Iwasaki et al., 1988a); mineral mix (Iwasaki et al., 1988a); and vitamin mix (Yu et al., 1982). In each of these studies, the life-extending effect of CR was the same whatever particular dietary component was not restricted; i.e., the determining factor was the extent to which total caloric intake was reduced, rather than the composition of the diet.

It has been suggested that the life-prolonging action of CR might be due to a decreased intake of adventitious toxic dietary contaminants rather than a decreased intake of calories. This is not a likely possibility. The studies with semisynthetic diets, in which individual dietary components were manipulated as well as the fact that many different food sources have been used by investigators strongly preclude a reduction in toxic contaminants as a factor involved in the life extension (Masoro, 1988).

The many studies discussed above lead to the conclusion that reducing caloric intake is by far the major, if not the sole, dietary factor responsible for CR's anti-aging and life-prolonging actions. The role of other dietary factors is minor at most. Nevertheless, manipulating *ad libitum* diets, such as altering the level or source of protein and the level or source of carbohydrate, can have a small but significant effect on rat longevity (Iwasaki et al., 1988b; Murtagh-Mark et al., 1995; Yu et al., 1985).

References

Austad, S. N. (1989). Life extension by dietary restriction in the bowl and doily spider, *Frontinella pyramitela*. *Exp. Gerontol. 24*: 83–92.

Berg, B. N. & Simms, H. S. (1960). Nutrition and longevity in the rat. II. Longevity and the onset of disease with different levels of intake. *J. Nutrition 71*: 255–263.

Birren, J. E. (1996). History of gerontology. In: J. E. Birren (Ed.), *Encyclopedia of Gerontology*, Vol. 1 (pp. 655–665). San Diego: Academic Press.

Cefalu, W. T., Wagner, J. D., Wang, Z. Q., Bell-Farrow, A. D., Collins, J., Haskell, D., Bechtold, R. & Morgan, T. (1997). A study of caloric restriction and cardiovascular aging in cynomolgus monkeys *(Macaca fascicularis)*: A model for aging research. *J. Gerontol.: Biol. Sci. 52A*: B10–B19.

Comfort, A. (1963). Effect of delayed and resumed growth on the longevity of fish *(Lebistes reticulates* Peter) in captivity. *Gerontologia (Basel) 8*: 150–155.

Duffy, P. H., Feuers, R. J., Pipkin, J. L., & Hart, R. W. (1994). Effect of chronic caloric restriction: Physiological and behavioral response to alternate day feeding in old female mice. *Age 17*: 13–21.

Duffy, P. H., Seng, J. E., Lewis, S. M., Mayhugh, M. A., Hattan, D. O., Casciano, D. A., & Feuers, R. J. (2001). The effects of different levels of dietary restriction on aging and survival. *Aging Clin. Exper. Res. 13*: 263–272.

Fanestil, D. D. & Barrows, C. H., Jr. (1965). Aging in the rotifer. *J. Gerontol. 20*: 462–469.

Goodrick, C. L., Ingram, D. K., Reynolds, M. A., & Freeman, J. R. (1982). Effects of intermittent feeding upon growth and life span in rats. *Gerontology 28*: 233–241.

Hansen, B. C. & Bodkin, N. L. (1993). Primary prevention of diabetes mellitus by prevention of obesity in monkeys. *Diabetes 42*: 1809–1814.

Harrison, D. E. & Archer, J. R. (1987). Genetic differences in effects of food restriction on aging mice. *J. Nutrition 117*: 376–382.

Hekimi, L. S., Benard, C., Branicky, R., Burgess, J., Hihi, A. K., & Rea, S. (2001). Why only time will tell. *Mech. Ageing Dev. 122*: 571–594.

Ingram, D. K., Cutler, R. G., Weindruch, R., Renquist, D. M., Knapka, J. J., April, M., Belcher, C. T., Clark, M. A., Hatcherson, C. D., Marriott, B. M., & Roth, G. S. (1990). Dietary restriction and aging: The initiation of a primate study. *J. Gerontol.: Biol. Sci. 45*: B148–B163.

Iwasaki, K, Gleiser, C. A., Masoro, E. J., McMahan, C. A., Seo, E. & Yu, B. P. (1988a). Influence of the restriction of individual dietary components on longevity and age-related disease of Fischer rats: The fat component and the mineral component. *J. Gerontol.: Biol. Sci. 43*: B13–B21.

Iwasaki, K, Gleiser, C. A., Masoro, E. J., McMahan, C. A., Seo, E. J. & Yu, B. P. (1988b). The influence of dietary protein source on longevity and age-related disease processes of Fischer rats. *J. Gerontol.: Biol. Sci. 43*: B5–B12.

Kealy, R. D., Lawler, D. F., Ballam, J. M., Mantz, S. L., Biery, D. N., Greeley, E. H., Lust, G., Segre, M., Smith, G. K., & Stowe, H. D. (2002). Effects of diet restriction on life span and age-related changes in dogs. *J. Am. Vet. Med. Assoc. 220*: 1315–1320.

Kemnitz, J. W., Weindruch, R., Roecker, E. B., Crawford, K., Kaufman, P. L., Ershler, W. B. (1993). Dietary restriction of adult male rhesus monkeys: Design, methodology, and preliminary findings from the first year of study. *J. Gerontol.: Biol. Sci. 48*: B17–B26.

Kirk, K. L. (2001). Dietary restriction and aging: Comparative tests of evolutionary hypotheses. *J. Gerontol.: Biol. Sci. 56A*: B123–B129.

Klass, M. R. (1977). Aging in the nematode *Caenorhabditis elegans*: major biological and environmental factors influencing life span. *Mech. Ageing Dev. 6*: 413–429.

Lynch, M. & Ennis, R. (1983). Resource availability, maternal effects, longevity. *Exp. Gerontol. 18*: 147–165.

Masoro, E. J. (1988). Food restriction in rodents: An evaluation of its role in the study of aging. *J. Gerontol.: Biol. Sci. 43*: B57–B64.

Masoro, E. J. (1990). Assessment of nutritional components in prolongation of life and health by diet. *Proc. Soc. Exp. Biol. Med. 193*: 31–34.

Masoro, E. J. (1993). Dietary restriction and aging. *J. Am. Geriatr. Soc. 41*: 994–999.

Masoro, E. J. (1999). *Challenges of biological aging*. New York: Springer.

Masoro, E. J., Iwasaki, K., Gleiser, C. A., McMahan, C. A., Seo, E., & Yu, B. P. (1989). Dietary modulation of the progression of nephropathy in aging rats: An evaluation of the importance of protein. *Am. J. Clin. Nutr. 49*: 1217–1227.

McCay, C. M. (1942). Chemical aspects of ageing and the effect of diet upon ageing. In: E. V. Cowdry (Ed.), *Problems of ageing*, 2nd ed. (pp. 680–727). Baltimore: Williams & Wilkens.

McCay, C. M., Crowell, M. F., & Maynard, L. A. (1935), The effect of retarded growth upon the length of life and upon the ultimate body size. *J. Nutrition 10*: 63–79.

McCay, C. M., Dilley, W. E., & Crowell, M. F. (1929). Growth rates of brook trout reared upon purified rations, upon dry skim milk diets and upon feed combinations of cereal grains. *J. Nutrition 1*: 233–246.

McCay, C. M., Maynard, L. A., Sperling, G., & Barnes, L. L. (1939). Retarded growth, lifespan, ultimate body size, and age changes in the albino rat after feeding diets restricted in calories. *J. Nutrition 18*: 1–13.

Merry, R. L. & Holehan, A. M. (1985). *In vivo* DNA synthesis in dietary restricted long-lived rats. *Exp. Gerontol. 20*: 15–28.

Murtagh-Mark, C. M., Reiser, K. M., Harris, Jr., R., & McDonald, R. B. (1995). Source of dietary carbohydrate affects life span of Fischer 344 rats independent of caloric restriction. *J. Gerontol.: Biol. Sci. 50A*: B148–B154.

Osborne, T. B. & Mendel, L. B. (1915). The resumption of growth after a long continued failure to grow. *J. Biol. Chem. 23*: 435–454.

Osborne, T. B., Mendel, L. B., & Ferry, E. L. (1917). The effect of retardation of growth upon the breeding period and duration of life in rats. *Science 45*: 294–295.

Robertson, T. B. & Ray, L. A. (1920). On the growth of relatively long lived compared with that of relatively short lived animals. *J. Biol. Chem. 42*: 71–107.

Robine, J.-M., Vaupel, J. W., Jeune, B., & Allard, M. (Eds.) (1997). *Longevity: To The Limits and Beyond.* Berlin: Springer Verlag.

Ross, M. H. (1961). Length of life and nutrition in the rat. *J. Nutrition 75*: 197–210.

Ross, M. H. & Bras, G. (1965). Tumor incidence patterns and nutrition in the rat. *J. Nutrition 87*: 245–260.

Roth, G. S., Ingram, D. R., & Lane, M. A. (1999). Caloric restriction in primates: Will it work and how will we know? *J. Am. Geriatr. Soc. 47*: 896–903.

Rudzinska, M. A. (1951). The influence of amount of food on the reproduction rate and longevity of a suctorian *(Tokophrya infusionum).* *Science 113*: 10–11.

Sawada, M. & Carlson, J. C. (1987). Association between lipid peroxidation and life-modifying factors in rotifers. *J. Gerontol. 42*: 451–456.

Saxton, J. A., Jr. & Kimball, G. C. (1941). Relation to nephrosis and other diseases of albino rat to age and to modification of diet. *Arch. Path. 32*: 951–965.

Saxton, J. A. Jr., Boon, M. C., & Furth, J. (1944). Observations on the inhibition of development of spontaneous leukemia in mice by underfeeding. *Cancer Res. 4*: 401–409.

Semsei, I (2000). On nature of aging. *Mech. Ageing Dev. 117*: 93–108.

Stuchlikova, E., Juricova-Horakova, M. & Deyl, Z. (1975). New aspects of dietary effect of life prolongation in rodents. What is the role of obesity in aging? *Exp. Gerontol. 10*: 141–144.

Vanfleteren, J. R. & Braeckman, B. P. (1999). Mechanism of life span determination in *Caenorhabditis elegans. Neurobiol. Aging 20*: 487–502.

Velthuis, E. J. M., van der Berg, H., Schaafsmor, G., Hendriks, H. F. J. & Brouwer, A. (1994). Energy restriction, A useful intervention to retard human aging? *Eur. J. Clin. Nutr. 48*: 138–148.

Verdery, R. & Walford, R. L. (1998). Changes in plasma lipids and lipoproteins in humans during 2-year period of dietary restriction in Biosphere 2. *Arch. Intern. Med. 158*: 900–906.

Walford, R. L., Harris, S. B., & Gunion, M. W. (1992). The calorically restricted low-fat nutrient-dense diet in Biosphere 2 significantly lowers blood glucose, total leukocyte count, cholesterol, and blood pressure. *Proc. Natl. Acad. Sci. USA 89*: 11533–11537.

Walford, R. L., Mock, D., Verdery, R., & MacCallum, T. (2002). Caloric restriction in Biosphere 2: Alterations in physiologic, hematologic, hormonal, and biochemical parameters in humans restricted for a 2–year period. *J. Gerontol. Biol.: Sci. 57A*: B211–B224.

Weindruch, R. (1996). The retardation of aging by caloric restriction: Studies in rodents and primates. *Tox. Path. 24*: 742–745.

Weindruch, R. & Walford, R. L. (1988). *The Retardation of Aging and Disease by Dietary Restriction*. Springfield, IL: Charles C. Thomas.

Weindruch, R., Walford, R. L., Fligiel, S. & Guthrie, D. (1986). The retardation of aging in mice by dietary restriction: longevity, cancer, immunity and lifetime energy intake. *J. Nutrition 116*: 641–654.

Yu, B. P., Masoro, E. J., & McMahan, C. A. (1985). Nutritional influences on aging of Fischer 344 rats: Physical, metabolic, and longevity characteristics. *J. Gerontol. 40*: 657–670.

Yu, B. P., Masoro, E. J., Murata, I., Bertrand, H. A., & Lynd, F. T. (1982). Life span study of SPF Fischer 344 male rats fed *ad libitum* or restricted diets: Longevity, growth, lean body mass, and disease. *J. Gerontol. 37*: 130–141.

CHAPTER 2
Mortality characteristics

Contents

The mortality characteristics of populations have been, and most probably will continue to be, widely used in the study of aging. Data, relating to calendar age and mortality of a population or a cohort, are collected in what are termed life tables. There are two kinds of such tables: period life tables comprised of data on an entire population at a point in time, and cohort life tables which provide data on the entire existence of a cohort. It is the cohort life tables that are utilized in the analyses of CR studies. In these tables, age intervals covering the existence of the cohort are chosen by the compiler of the table; in rat and mouse studies, monthly age intervals are usually chosen, for example, 0–1 month of age, 1–2 months of age, and so on until the last member of the cohort is dead. For each age interval, the following information relative to mortality is determined: the number of animals or people in the cohort alive at the beginning of the interval; the number that died during the interval; and the age-specific death rate (i.e., the percentage of the cohort alive at the start of an age interval that dies during that interval). Ross (1961) reported that in the age range of 6–24 months, the age-specific death rate for rats on a CR regimen was significantly lower than that of *ad libitum*-fed control rats.

Entire life tables are rarely reported in gerontological publications because they contain such a mass of data that they would be difficult to readily comprehend, as well as expensive to print. Rather, selected data

RESEARCH PROFILES IN AGING
VOLUME 1 ISSN 1567-7184

are used, along with easily comprehended graphic illustrations based on the life tables.

Survival curves and related data

Survival curves are frequently used to create graphic presentations of life table mortality data. Figure 2-1 is an example of survival curves obtained in the analysis of data from a CR study on male F344 rats, which was carried out in our laboratory (Yu et al., 1985). The *y* axis denotes the percentage of the rat cohort that is alive, and the *x* axis denotes the age (in months) of the rats. Groups A (115 rats) and R (115 rats) are from a cohort of 230 male F344 rats studied from 1975 to 1979. The rats in both groups were fed *ad libitum* until 6 weeks of age; then the rats in Group R were restricted to 40% fewer calories than the mean intake of Group A. Groups 1 (40 rats) and 2 (40 rats) are a cohort of 80 rats studied from 1980 to 1984; these groups were maintained on the

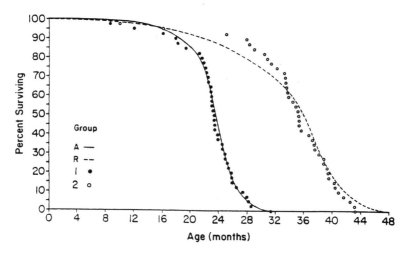

Figure 2-1. Survival curves for male F344 rats fed *ad libitum* or restricted to 40% of the *ad libitum* intake. In the first study, survival curves were generated from 115 *ad libitum* fed rats (Group A) and 115 restricted rats (Group R). In the second study, conducted 4 years later, the survival curves were generated from 40 *ad libitum*-fed rats (Group 1) and 40 restricted rats (Group 2). The survival curves of Groups A and R differed significantly, as did those of Groups 1 and 2 ($P > 0.01$). The survival curves of Groups A and 1 did not differ significantly nor did those of Groups R and 2. (From *J. Gerontol. 40*: 657–670, 1985; Copyright © The Gerontological Society of America. Reproduced by permission of the publisher.)

same protocol as rats in Group A and Group R, respectively. A cursory glance shows that the curves derived from Groups R and 2 are markedly to the right of those from Groups A and 1. CR affects the survival curve of female F344 rats in a similar way (Thurman et al., 1994).

Clearly, these studies show that CR increases both the median length of life and the maximum length of life (the age at death of the longest-lived animal) of the population. However, an increase in median length of life does not necessarily mean that aging processes have been slowed; prevention of premature death would increase the median length of life even if the aging processes were not affected. In fact, the shapes of the curves in Figure 2-1 indicate that prevention of premature death is not a likely factor in the effect of CR on the survival curves. It is important to note that an increase in the maximum length of life of a population has long been viewed as strong evidence of a decrease in the rate of aging. But, it is obviously risky to draw any conclusion about the rate of aging with information from a single animal (the last animal to die). Indeed, the reliability of data on the maximum length of life of populations has been questioned on several grounds (Gavrilov & Gavrilova, 1990). There are two measurements that are considered more reliable indicators of the rate of aging, because they are not based on the death of a single animal. One is the 10th percentile survival time (Yu et al., 1985), and the other is average age at death of the longest-lived 10% of the cohort of animals; the latter is a measure that Lewis et al. (1999) have termed the maximal survival (a variation of the dictionary definition of maximal). As is evident from Figure 2-1, CR markedly increases both the 10th percentile survival time and the maximal survival. The other striking aspect of Figure 2-1 is the similarity of survival curves A and 1 and of survival curves R and 2. That curves 1 and 2 are not as smoothly shaped as those of A and R relates to the fact that 115 rats were used for each dietary group in the 1975–79 study and only 40 in the 1980–1984 investigation. Overall, Figure 2-1 shows that CR's effect on longevity is both robust and reproducible.

Similar findings have been noted in studies on both genders of many different strains of rats and mice. Bertrand et al. (1999) have tabulated such information from studies of six rat strains (F344, BN, BN × F344 F1, Wistar, Lobund-Wistar, and Sprague-Dawley) and nine mouse strains (C57BL/6, Balb/c, MRl/lpr, DBA/2f, B10C3F1, C3B10RF1, B6C3F1, NZB × NZW, and Emory).

Results of a study conducted at the National Center for Toxicological Research (NCTR) in Jefferson, AR, are particularly informative since CR was studied simultaneously in the same laboratory on both genders of three rat strains (F344, BN, and F344 × BN F1) and four mouse strains

Table 2-1
Summary of effects of CR on longevity of rat strains in NCTR study

Rat strain	Gender	Percent increase due to CR in	
		Median survival	Maximal survival
F344	Male	20	28
	Female	17	13
BN	Male	16	20
	Female	26	20
F344 × BNF1	Male	20	22
	Female	36	35

Note: This table was generated from data in the report of Lewis et al. (1999).

(C57Bl/6, B6C3F1, DBA, and B6D2F1) under as nearly identical environmental conditions as possible. Accounts of the survival character-istics of the animals in this study have been published (Lewis et al., 1999; Turturro et al., 1999). The restriction of calories was started gradually at 14 weeks of age, reaching 40% of the intake of the control (*ad libitum* fed) animals by 16 weeks of age; and the 40% level of restriction was maintained for the remainder of life of the rats and mice. The effects of CR on the longevity of the rat strains are summarized in Table 2-1, and the mouse strains in Table 2-2. CR increased longevity in both genders of all rat and mouse strains in the NCTR study, though the magnitude of increase varied with strain and gender.

Figure 2-2 is a graph published by Feuers et al. (1993) in which the severity of caloric restriction is related to average and maximum life span of a population of female C3B10RF1 mice. The data used to generate this graph were from a study by Weindruch et al. (1986), in which, groups of these mice were fed 125 or 85 or 50 or 40 kcal/week. Over this range of caloric intake, it is clear that the lower the intake, the longer the life span. Of course, there would be a lower limit at which caloric intake would be less than needed to sustain life, but it is clear that it can be markedly reduced before that limit is reached.

Holloszy (1997) studied the interaction of exercise and CR on the survival characteristics of male Long-Evans rats. It was found that such interaction had neither additive nor synergistic effects on survival.

Gompertzian analyses

In the 19th century, Benjamin Gompertz pointed out that the age-specific mortality of adult human populations is an exponentially

Table 2-2
Summary of effects of CR on longevity of mouse strains in NCTR study

Mouse strain	Gender	Percent increase due to CR in	
		Median survival	Maximal survival
B6C3F1	Male	36	21
	Female	36	29
C57BL/6	Male	15	19
	Female	25	18
DBA/2	Male	18	12
	Female	53	12
B6D2F1	Male	36	18
	Female	33	28

Note: This table was generated from data in the report of Lewis et al. (1999).

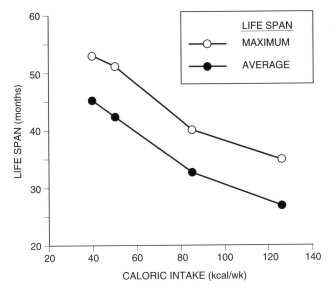

Figure 2-2. The relation between caloric intake and length of life of female C3B10RF1 mice. CR started at 3 weeks of age. Maximum life span is defined as the mean length of life of the longest lived decile at each level of caloric intake; i.e., it is what Lewis et al. (1999) term the maximal survival. (From Feuers et al., 1993.)

increasing function of calendar age. While this relationship has been found to be valid for many species through much of the adult life span, it ceases to be true, at least in some species—including humans—at advanced ages (Wilmouth & Horiuchi, 1999). The rate of the exponential increase in age-specific mortality has been widely used as an index of the rate of population aging (Finch, 1990). The amount of time elapsed before the age-specific mortality rate doubles, called the Mortality Rate Doubling Time (MRDT), relates inversely to the rate of increase in the age-specific death rate. Thus, the MRDT provides an easily determined inverse index of the rate of population aging.

Holehan and Merry (1986) used data from studies in their laboratory (Merry & Holehan, 1985) and our laboratory (Yu et al., 1982) to calculate the effects of CR on the MRDT. In their study, the control rats were fed *ad libitum* and the rats on CR were fed the amount of food needed to maintain a body weight 50% that of the control rats; the MRDT of the control rats was 102 days and that of the CR rats was 203 days. In our study, the rats on CR were given 40% fewer calories than the control rats; the MRDT of the control rats was 104 days and that of the CR rats was 189 days. Obviously, the effect of CR on the MRDT was remarkably similar in these studies, though they differed markedly in the method of achieving CR. Pletcher et al. (2000) developed a method of assessing Gompertzian and related analyses of CR's effect on mortality rate at all ages, as well as its effect on the increase in mortality with advancing age; they used the method to analyze our data on male F344 rats, and concluded that CR markedly slows the rate of population aging in the male gender of this rat strain. This method of Pletcher and associates should prove invaluable in analyzing the extensive database on both genders of a spectrum of mouse and rat strains already in the literature as well as findings generated by future studies.

Age of initiation and duration of CR

Because McCay and colleagues emphasized the role of growth retardation in life extension, most CR studies have been initiated in young rats and mice. However, it has become clear that CR need not begin with young growing rodents to markedly affect longevity.

In our laboratory (Yu et al., 1985), we randomly assigned members of a cohort of male F344 rats to the following dietary regimens: Group 1, fed *ad libitum* throughout life; Group 2, fed *ad libitum* until 6 weeks of age and on the CR regime for the rest of life; Group 3, fed *ad libitum* until 6 weeks of age, the CR regime from 6 to 26 weeks of age, and then again *ad libitum* for the rest of life; Group 4, fed *ad libitum* until 6 months of age and the CR regime for the rest of life. In Groups 2, 3, and 4, CR involved a 40% reduction in intake of calories below that of the *ad libitum* fed rats. The longevity findings are summarized in Table 2-3. Each of the CR dietary regimens increased both median and 10th percentile survival.

Table 2-3
Effects of age of initiation and duration of CR on longevity of rats

Group	Time on CR	Median survival, days	10th percentile survival, days
1	None	701	822
2	from 6 weeks of age	1057	1226
3	6 weeks to 26 weeks of age	808	918
4	from 6 months of age	941	1177

Note: This table was generated from data in the report of Yu et al. (1985).

It is striking that as little as 3½ months of restriction (Group 3) increased both median survival and 10th percentile survival by about 100 days. It is even more striking that CR started at 6 months of age was almost as effective as CR started at 6 weeks of age in increasing 10th percentile survival. At 6 weeks of age, male F344 rats are growing rapidly, but that is not the case at 6 months of age when almost full skeletal length has been reached. This study clearly demonstrates that retardation of growth and development does not play a major role in the life-extending action of CR.

Indeed, Weindruch and Walford (1982) showed that CR can be initiated in mice at middle age and still achieve a significant life-extending action. They gradually established CR (about 30% below the caloric intake of control mice) in 12-month-old mice of the B10C3F1 and C57BL/6 strains; CR increased the maximal life span of these mice by 10–20%. However, with regard to the rat, studies have shown that CR does not increase longevity when initiated at advanced ages. CR begun at 18 months of age in Long-Evans rats did not affect life span (Lipman et al., 1995) nor did it when initiated at 18 or 26 months of age in F344 × BNF1 rats (Lipman et al., 1998). Moreover, it is interesting to note that there was a decrease in length of life when CR was started in Sprague-Dawley rats at 300 days of age or older (Ross, 1978).

Temporal pattern of food intake

Although *ad libitum*-fed rodents eat primarily during the dark period of the light–dark cycle, they continue to eat during the light period, albeit at a lower rate; i.e., they are nibblers rather than meal eaters. However, the two CR procedures (a single daily meal and every-other-day feeding) involve a meal-eating pattern followed by many hours of fasting.

Table 2-4
Effects of eating pattern on longevity effects of CR

Dietary regimen	Median survival, days	10th percentile survival, days
Ad Libitum Fed	768 (744–790)	868 (849–982)
CR (1 meal/day)	1035 (966–1075)	1195 (1117–1244)
CR (2 meals/day)	989 (918–1044)	1167 (1136–1266)

Note: Survival data in parentheses are 95% confidence intervals; this table was generated from data in the report of Masoro et al. (1995).

The question thus arose as to whether the life-extending effect of CR was due to reduced caloric intake or periods of fasting. Masoro et al. (1995) addressed this question by comparing the effects of CR when food was given to one group in a single daily meal at 15:00 h and to another group in two daily meals (07:00 h and 15:00 h). Both regimens increased median survival and 10th percentile survival to the same extent (Table 2-4). Therefore, it was concluded that reduction of energy intake rather than temporal pattern of food intake is responsible for the life-extending action of CR.

References

Bertrand, H. A., Herlihy, J. T., Ikeno, Y., & Yu, B. P. (1999). Dietary restriction. In: B. P. Yu, (Ed.), *Methods in aging research* (pp. 271–300). Boca Raton, FL: CRC Press.

Feuers, R. J., Weindruch, R., & Hart, R. W. (1993). Caloric restriction, aging, and antioxidant enzymes. *Mutation Res. 295*: 191–200.

Finch, C. E. (1990). *Longevity, Senescence, and the Genome*. Chicago: University of Chicago Press.

Gavrilov, L. A. & Gavrilova, N. S (1990) *The Biology of Life: A Quantitative Approach*. Chur, Switzerland: Harwood Academic Publications.

Holehan, A. M. & Merry, B. J. (1986). The experimental manipulation of ageing by diet. *Biol. Rev. 61:* 329–369.

Holloszy, J. O. (1997). Mortality rate and longevity of food restricted male rats: a reevaluation. *J. Appl. Physiol. 82:* 399–403.

Lewis, S. M., Leard, B. L., Turturro, A., & Hart, R. W. (1999). Long-term housing of rodents under specific pathogen-free barrier conditions. In: B. P. Yu (Ed.), *Methods in Aging Research* (pp. 217–235). Boca Raton, FL: CRC Press.

Lipman, R. D., Smith, D. E., Blumberg, J. B., & Bronson, R. T. (1998). Effects of caloric restriction or augmentation in adult rats: Longevity and lesion biomarkers of aging. *Aging Clin. Exp. Res. 10:* 463–470.

Lipman, R. D., Smith, D. E., Bronson, R. R., & Blumberg, J. (1995). Is late-life caloric restriction beneficial? *Aging Clin. Exp. Res. 7:* 126–129.

Masoro, E. J., Shimokawa, I., Higami, Y., McMahan, C. A., & Yu, B. P. (1995). Temporal pattern of food intake not a factor in the retardation of aging processes by dietary restriction. *J. Gerontol.: Biol. Sci. 50A:* B48–B53.

Merry, B. J. & Holehan, A. M. (1985). *In vivo* DNA synthesis in dietary restricted long-lived rats. *Exp. Gerontol. 20:* 15–28.

Pletcher, S. D., Khazaeli, A.A., & Curtsinger, J. W. (2000). Why do life spans differ? Partitioning mean longevity differences in terms of age-specific mortality parameters. *J. Gerontol.: Biol. Sci. 55A:* B381–B389.

Ross, M. H. (1961). Length of life and nutrition in the rat. *J. Nutrition 75:* 197–210.

Ross, M. H. (1978). Nutrition regulation of longevity. In: Behnke, A., Finch, C., & Moment, G. B. (Eds.), *Biology of Aging* (pp. 175–189). New York, Plenum Press.

Thurman, J. D., Bucci, T. J., Hart R. W., & Turturro, A. (1994). Survival, body weight, and spontaneous neoplasms in ad libitum-fed and food-restricted rats. *Tox. Path. 22:* 1–9.

Turturro, A., Witt, W. W., Lewis, S., Hass, B., Lipman, R. D., & Hart, R. W. (1999). Growth curves and survival characteristics of the animals used in the Biomarkers of Aging Program. *J. Gerontol.: Biol. Sci. 54A:* B492–B501.

Weindruch, R. & Walford, R. L. (1982). Dietary restriction in mice beginning at 1 year of age: effects on lifespan and spontaneous cancer incidence. *Science 215:* 1415–1418.

Weindruch, R., Walford, R. L., Fligel, S., & Guthrie, D. (1986). The retardation of aging in mice by dietary restriction: longevity, cancer, immunity, and lifetime energy intake. *J. Nutrition 116:* 642–654.

Wilmouth, J. R. & Horiuchi, S. (1999). Do the oldest old grow old more slowly? In: J.-M. Robine, B. Forrette, C. Franceschi, & M. Allard (Eds.), *The Paradoxes of Longevity* (pp. 35–60). Berlin: Springer Verlag.

Yu, B. P., Masoro, E. J., & McMahan, C. A. (1985). Life span study of SPF Fischer 344 male rats. I. Physical, metabolic, and longevity characteristics. *J. Gerontol. 40:* 657–670.

Yu, B. P., Masoro, E. J., Murata, I., Bertrand, H. A., & Lynd, F. T. (1982). Life span study of SPF Fischer 344 male rats fed *ad libitum* or restricted diets: Longevity, growth, lean body mass, and disease. *J. Gerontol. 37:* 130–141.

CHAPTER 3

Molecular and cellular biology

Contents

Any intervention that slows the aging processes probably does so by affecting fundamental molecular and cellular functions. CR influences many cellular processes, causing some functions to change almost immediately and slowing age-associated changes in others. One or more of these cellular and/or molecular effects is (are) likely to play key roles in CR's anti-aging and life-extending actions.

Stability of the nuclear genome

It has long been theorized that nuclear genomic instability is the basis of senescence (Failla, 1958; Szilard, 1959). Indeed, genomic DNA damage and mutations are known to increase with age (Bohr & Anson, 1995). CR has been associated with the attenuation of certain types of age-associated DNA damage (Haley-Zitlin & Richardson, 1993). It protects against DNA oxidative damage in rats and mice (Chung et al., 1992; Djuric et al., 1992; Kaneko et al., 1997; Sohal et al., 1994a). Figure 3-1 illustrates the marked ability of CR to attenuate the age-associated increase in oxidatively damaged DNA in the tissues of male C57BL/6 mice (Sohal et al., 1994a). Based on the hypoxanthine phosphoribosyl

RESEARCH PROFILES IN AGING
VOLUME 1 ISSN 1567-7184

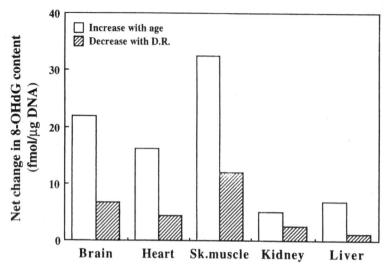

Figure 3-1. Effect of age on 8-OHdG content of the DNA in the tissues of male C57BL/6 mice and CR's ability to decrease its content. The open bars denote the difference in content between 8 and 27 month old mice. The hatched bars denote the decreased content in the tissues of 15-month-old mice that have been on long-term CR compared to mice of the same age on an *ad libitum* diet. The authors refer to CR as D.R. (From Sohal et al., 1994a.)

Table 3-1
Geometric mean of mutation frequencies in lymphocytes of *ad libitum* (AL)-fed and CR mice

	Mutation frequency $\times 10^{-6}$	
Diet group	Age, 6 months	Age, 12 months
AL	10.1 (8.0–12.7)	35.6 (28.5–44.5)
CR	8.1 (5.6–11.7)	8.0 (5.3–12.2)

Note: Numbers in parentheses are 95% confidence limits. (From Dempsey et al., 1993.)

transferase *(hprt)* gene locus test, CR has also been found to slow the age-associated increase in genomic mutations in lymphocytes of mice and rats (Aidoo et al., 1999; Casciano et al., 1996; Dempsey et al., 1993). Table 3-1 summarizes the findings of Dempsey et al. (1993) on mutation frequencies in the lymphocytes of 6- and 12-month-old female BALB/c mice fed *ad libitum* or on a CR regimen.

The age-associated increase in mutations and DNA damage could be the result, in part or in whole, of a decreased ability to repair DNA.

Weraachakul et al. (1989) used unscheduled DNA synthesis (UDS) in cells exposed to ultraviolet light as a measure of DNA repair ability; and they found that hepatocytes and kidney cells isolated from male F344 rats exhibited a slowing of repair as the age of the rat increased. At all the ages studied, cells from rats on CR had a higher rate of UDS than *ad libitum*-fed rats of the same age. Using UDS and various rat and mouse cell types, other investigators have also concluded that CR enhances DNA repair (Licastro et al., 1988; Lipman et al., 1989; Srivastava & Busbee, 1992; Tilley et al., 1992). Also, undernutrition has been found to increase DNA repair in human peripheral lymphocytes (Rao et al., 1996).

Unfortunately, UDS is such a crude assay that it neither directly nor selectively measures repair of a specific type of DNA damage. It does, however, provide a measure of overall genome repair. In mammalian cells, the removal of ultraviolet light-induced DNA damage involves the nucleotide excision repair pathway, which preferentially repairs transcriptionally active genes; the transcribed strand is more rapidly repaired than the nontranscribed strand. The rapid repair of transcribed strands is called strand-specific repair, or transcription-coupled repair, or preferential repair, while the repair of nontranscribed DNA is referred to as global repair or bulk repair (Friedberg, 1996). Guo et al. (1998a) studied the effect of age and CR on nucleotide excision repair, comparing the transcribed albumin gene of isolated hepatocytes and nontranscribed DNA of these cells. The rate of repair of the transcribed strand of albumin DNA was 40% less in hepatocytes from 24-month-old rats than 6-month-old rats, and CR prevented this age-associated decrease. While there was also a decrease with age in global repair, CR did not prevent this decrease. Guo et al. (1998b) studied the repair of the phosphoenolpyruvate carboxykinase (PEPCK) gene in hepatocytes of young and old rats. When the PEPCK gene was not being transcribed, the slow rate of repair was not affected by age. The addition of cAMP rapidly induced PEPCK transcription, markedly increasing the rate of repair of the transcribed strand in hepatocytes from young *ad libitum*-fed rats, though not those from old rats. However, lifelong CR resulted in a similar increase in the repair of the transcribed strand in hepatocytes from both old and young rats.

Gene expression

Changes in gene expression have also long been viewed as a factor underlying senescence (Vijg, 1996). Most of the studies have used the level of the mRNA transcript of a gene as an index of the expression of that gene (Van Remmen et al., 1995). Although this index provides valuable information, it is the protein encoded by the DNA of a gene that is usually the final product of gene expression, and the mRNA level does not always correlate with the rate of generation of that protein. Nor, for that matter, does the level of the mRNA transcript always correlate with the rate of transcription of the gene. It is important to keep these caveats in mind when evaluating the meaning of mRNA levels.

With increasing age, the mRNA transcript level increases for some genes, decreases for others, and does not change for many; the findings of many studies on the effect of age on mRNA transcript levels are summarized in Tables 9.3 through 9.6 in a review article by Van Remmen et al. (1995). CR in rats and mice has been found to blunt the age change in the level of many mRNA transcripts, affecting both those that increase and those that decrease with increasing age. However, it is difficult to draw general conclusions about the effects of CR from these data for the following reasons: The mRNA levels have been measured for only a few genes; mixtures of cell types rather than single cell types have been studied; and few detailed life-span studies have been done.

However, rapidly developing technology for the study of cellular and molecular biology should soon rectify this situation. The advent of high-density array analyses of the level of mRNA transcripts is already addressing the first issue (Ramsay, 1998). Twofold changes in the levels of mRNA transcripts can be detected for hundreds, even thousands, of genes screened simultaneously.

Weindruch and his colleagues at the University of Wisconsin were the first to use this technique to study the effects of aging and CR. These investigators analyzed the levels of mRNA transcripts of 6347 genes in the gastrocnemius muscles of 5- and 30-month-old mice. Of the genes screened, 58 showed more than a twofold increase in mRNA levels and 55 more than a twofold decrease in mRNA levels in the muscles of 30-month-old mice compared to 5-month-old mice (Lee et al., 1999). Most of the age-associated changes in the levels of mRNA transcripts were partially or completely prevented by CR. Of the transcripts that decreased with age, 13% coded for proteins involved in energy metabolism; and of those that increased with age, 16% coded for proteins involved in stress responses. However, like most pioneering research, this study has some deficiencies. There was no statistical analysis to determine if the changes were significant. Furthermore, no attempt was made to validate the array analyses by an independent method, such as Northern or Reverse Transcription Polymerase Chain Reaction (RT-PCR) analyses. This array technology was also used to study the effect of CR on gene expression in the vastus lateralis muscle of 20-year-old rhesus monkeys (Kayo et al., 2001); CR resulted in an upregulation of cytoskeletal protein encoding genes and a decrease in the expression of genes involved in mitochondrial bioenergetics. Based on these muscle studies, Weindruch et al. (2001) suggest that transcriptional patterns in muscle indicate that CR retards the aging processes by causing a metabolic shift toward increased protein turnover and decreased macromolecular damage. It is difficult to find the basis for this suggestion in the two muscle studies that have been published. Rather, it seems clear that further developments in informatics are required for reliable interpretation of these complex data sets.

Lee et al. (2000) extended this line of research to the effects of age and CR on the gene-expression profile of the neocortex and cerebellum of 5- and 30-month-old mice. Of the 6347 genes assessed, the expression of 63 genes was interpreted as increasing with age by at least 1.7-fold. In the neocortex, 20% of these genes were classified as inflammatory response genes, and 24% as stress response genes. In the cerebellum, 27% were classified as inflammatory response genes and 13% as stress response genes. Aging appeared to decrease the expression of 47 genes in the neocortex and 63 genes in the cerebellum; many of these genes coded for transcription factors or trophic functions. CR partially or completely prevented many of the perceived age changes in gene expression and appeared to selectively attenuate age changes in genes encoding inflammatory and stress responses. Some of these findings were confirmed by RT-PCR, but details of this aspect of the study were not included in the published communication of Lee et al.

Weindruch et al. (2002) reviewed their findings on gene expression profiling of the skeletal muscle and the central nervous system to determine if there were general trends. They concluded that aging influences the expression of genes in four major physiologic classes: stress response, biosynthesis, protein metabolism, and energy metabolism. CR appeared to completely or partially suppress the age changes in the expression of 84% of them.

Other investigators have used high-density array gene expression technology to study the effects of aging and CR on the gene expression profile of liver. Han et al. (2000) assessed 588 genes from the livers of 3- and 24-month-old *ad libitum*-fed and CR mice. They independently verified the array findings by RT-PCR, and then carried out a statistical analysis. The levels of only six of the mRNA transcripts differed in the 3- and 24-month-old mice, and CR attenuated just one of these differences. Moreover, CR affected only four of the 582 transcripts that did not exhibit a change with age. Using the array technology for the study of gene expression in the liver of mice, Cao et al. (2001) reported age-related increases in the expression of genes associated with inflammation, cellular stress, and fibrosis and decreases in the expression of genes associated with apoptosis, xenobiotic metabolism, cell cycling, and DNA replication; long-term CR as well as short-term CR (duration of 4 weeks) reversed most of these age changes in hepatic gene expression. Teillet et al. (2002) used the array technology to study the liver of rats and found that CR does not uniquely influence genes with age-associated changes in expression, but rather produces a new profile of gene expression that includes genes related to lipid metabolism as well as those in the energetic pathways.

It is apparent that high-density array technology is in its infancy in regard to the study of the influence of aging and CR on gene expression. To date, only a small fraction of the genes in tissues of mixed cell types has been evaluated. However, this technology holds great promise, particularly if it becomes possible in the future to assess the expression of all genes of specific cell types. Also needed are improvements (1) in the technology that would permit the reliable assessment of much smaller differences than twofold, and (2) in informatics for interpretation of the findings.

A few CR studies have measured both the mRNA transcripts and the protein products. CR increases the mRNA transcript level and the rate of synthesis of α_{2v}-globulin in rat liver (Richardson et al., 1987). It also increases the levels of the mRNA transcripts for superoxide dismutase, catalase, and glutathione peroxidase, and their enzymatic activities in rat liver (Rao et al., 1990). CR causes a decrease in senescence marker protein and its mRNA level in rat liver (Chatterjee et al., 1989). In rat thyroid, it decreases calcitonin-gene-related-peptide and its mRNA level (Salih et al., 1993). Although aging does not alter the hepatic expression of glucose-regulated protein 78 (GRP78) in mice, CR decreases its expression (Spindler et al., 1990). It acts by destabilizing GRP78-mRNA, thereby decreasing its level in liver cells (Tillman et al., 1996). CR also decreases the expression of glucose-regulated protein 94 (GRP94) and its mRNA in mouse liver (Spindler et al., 1990).

Research on the influence of age and CR on the induction of heat-shock protein 70 (hsp70) in cells has yielded particularly penetrating insights. This protein, which is induced in cells by heat as well as other cellular stressors, protects cells from the damaging actions of stressors. With increasing age, there is a decrease in the ability of heat stress to promote the induction of hsp protein and its mRNA in rat hepatocytes. Heydari et al. (1993) found that CR increases the induction of hsp

Table 3-2
Induction of HSP 70 in hepatocytes from male F344 rats

Age of donor (months)	Protein synthesis		mRNA level	
	AL	CR	AL	CR
4–6	2.15 (0.38)	3.05 (0.27)	37.6 (3.8)	47.6 (5.5)
26–28	1.17 (0.33)	2.38 (0.32)	22.7 (4.1)	35.3 (2.2)

Note: Numbers in parentheses are Standard Errors. (From Heydari et al., 1993.)

protein and the level of its mRNA in hepatocytes of young and old rats (Table 3-2). These effects are due to an age-associated decrease in the induction of the transcription of the *hsp 70* gene in rat hepatocytes; CR increases the induction of *hsp 70* transcription in hepatocytes of rats of all ages (Heydari et al., 1995). A decreased ability of the transcription factor HSF1 to bind to its DNA binding site in rat hepatocytes appears to cause the age-associated attenuation of the induction of *hsp* transcription (Heydari et al., 2000). It is likely that CR enhances the transcription of the *hsp 70* gene by altering the molecular structure of the hepatic HSF1 protein. The induction of hsp 70 by heat stress decreases with increasing age in rat alveolar macrophages adhered to plastic; CR enhances induction in these macrophages from both young and old rats (Moore et al., 1998). While the heat-stress induction of proteins hsp 27, 70, and 90 in rat hypothalamus is attenuated with increasing age, CR blunts the extent of this attenuation (Aly et al., 1994). However, CR does not affect the induction of hsp 70 protein in rat spleen lymphocytes (Pahlavani et al., 1996).

Pahlavani et al. (1997) reported that in response to concanavalin A, there is an age-associated decrease in the induction of interleukin-2 (IL-2) activity and its mRNA levels in rat splenic T-cells; this also relates to the decreased DNA binding of a transcription factor, namely the nuclear factor of activated T-cells (NFAT). CR attenuates the age-associated decrease in the induction of IL-2 activity and increases the binding of NFAT to DNA. CR also blunts the age-associated decrease in the induction of c-fos, a component of the NFAT-protein complex.

The induction of the enzyme PEPCK in liver in response to fasting helps animals to cope with this challenge because of the enzyme's role in gluconeogenesis. Van Remmen and Ward (1998) found that the induction of this enzyme in response to fasting is lost in rats with advancing age

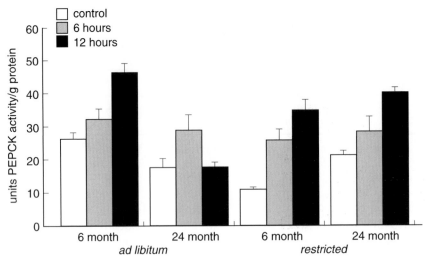

Figure 3-2. Effect of age and long-term CR on the ability of fasting to induce hepatic PEPCK activity in male 6- and 24-month-old F344 rats. The rats were fasted for 30 h and then refed for 24 h. Control refers to the activity at the end of the 24-h refeeding and 6 h and 12 h refer to the length of the fast following the refeeding. In the graph, CR is denoted by the word restricted. (From Van Remmen and Ward, 1998.)

and that CR prevents this age-associated loss (Figure 3-2). Indeed, CR influences hepatic gene expression so as to increase the enzymatic capacity for gluconeogenesis and the disposal of by-products of protein catabolism, as well as decreasing the enzymatic capacity for glycolysis (Dhahbi et al., 1999).

Studies on the effects of CR on gene expression are just beginning, but the limited information currently available indicates that such studies are likely to offer insight into the mechanisms underlying CR's anti-aging action. Gene expression appears to be influenced by CR so as to enable the organism to more effectively meet challenges ranging from fasting to oxidative stress; and it is the long-term effects of repeated occurrence of such challenges that may play a major role in the aging of organisms.

Mitochondrial function

Mitochondria play the central role in cellular energetics, and they are also involved in cell death (both apoptosis and necrosis) as well as cellular signaling pathways. Each mitochondrion has its own genome that codes

for key mitochondrial proteins. Many believe that alterations in mitochondria and their genomes are intimately involved in organismic aging
(Beckman & Ames, 1998).

Weindruch et al. (1980) reported that CR increases the respiratory
control index in isolated mouse mitochondria, which led them to suggest
that CR increases the efficiency of mitochondrial electron transport and
oxidative phosphorylation. Sohal et al. (1994b) carried out extensive
studies on isolated mitochondria from brain, heart, and kidney of 9-, 17-
and 23-month-old mice. CR prevents the age-associated increase in state
4 (resting) respiration that occurs in mitochondria from *ad libitum*-fed
mice. This *in vitro* study indicates that the rate of mitochondrial
generation of superoxide (Figure 3-3) and hydrogen peroxide (Figure 3-4)
increases with age and that CR lowers the rate of their generation.

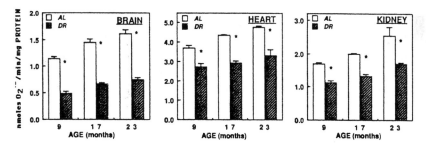

Figure 3-3. Rates of superoxide generation by submitochondrial particles from
mitochondria of 9-, 17-, and 23-month-old male C57BL/6 mice fed *ad libitum*
(AL) or maintained on a long-term CR diet (the authors designate CR as DR).
(From Sohal et al., 1994b.)

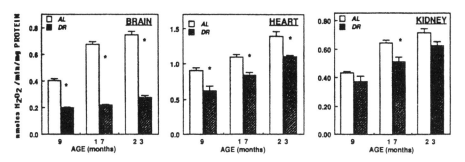

Figure 3-4. Rates of hydrogen peroxide release by mitochondria studied *in vitro*
from 9-, 17-, and 23-month-old male C57BL/6 mice fed *ad libitum* (AL) or
maintained on a long-term CR diet (the authors designate CR as DR). (From
Sohal et al., 1994b.)

Desai et al. (1996) found that CR opposes all age-associated changes in the mitochondrial electron transport chain (ETC) in mouse gastrocnemius muscle. Feuers (1998) has systematically studied the influence of CR and age on the components of the mitochondrial ETC in mouse gastrocnemius muscle. He found that with aging, there is a decrease in the amount of complexes I and III of the ETC chain and an increase in the K_m of complex III; in addition, the high-affinity binding sites of complex IV are lost. He suggests that these alterations underlie the age-associated increase in mitochondrial generation of reactive oxygen molecules, and that by attenuating these alterations in the ETC, CR decreases the formation of reactive oxygen molecules. Interestingly, Olgun et al. (2002) suggest that a decrease in the complex III/IV ratio contributes to CR's beneficial actions. Strikingly, CR blunts the age-associated increase of proton leak-dependent oxygen consumption in rat skeletal muscle mitochondria (Lal et al., 2001).

The investigations just discussed have yielded intriguing findings. However, our knowledge of the influence of CR and aging on the functioning of mitochondria in cellular energetics is still rudimentary. In the research to date, isolated mitochondria have been studied under *in vitro* conditions in which the levels of oxygen, substrates, and a host of other factors are very different from the levels in the cells of a living organism. Moreover, even in the *in vitro* setting, key information, such as the direct assessment of the effect of CR on oxidative phosphorylation, is lacking.

However, there is some evidence indicating that CR influences mitochondria in the intact organism in a beneficial way. CR reduces the level of mitochondrial lipid and DNA damage in the liver of old rats (Chung et al., 1992) as well as mitochondrial protein and DNA damage in the skeletal muscle of old mice (Lass et al., 1998). Also, CR attenuates the age-associated increase in mitochondrial DNA deletions in rat skeletal muscle and liver, but not in brain (Aspnes et al., 1997; Kang et al., 1998).

The findings of Kristal and Yu (1998) on the effect of CR on "permeability transition" (PT) in isolated rat mitochondria are particularly noteworthy, both in terms of magnitude and potential importance in regard to aging. CR markedly slows the induction of PT in response to 25 μM calcium or calcium plus t-butyl hydroperoxide in isolated mitochondria from 24-month-old rats.

Mitochondrial function is closely related to the "Oxidative Stress Theory of Aging" (Sohal & Dubey, 1994). This theory will be discussed in Chapter 6.

Membrane structure and function

Cellular membranes undergo changes with increasing age (Naeim & Walford, 1985), and it has been hypothesized that such changes play a major role in age associated cellular deterioration (Zs-Nagy, 1979). Indeed, many of the age changes in mitochondria just discussed may well be due to age-associated alterations in the membranes of these organelles. Pieri et al. (1990) proposed that the anti-aging action of CR may be due to the protection of the physiological properties of cellular membranes. Indeed, CR has been found to retard many of the age-associated changes in cellular membranes (Yu, 1995).

With increasing age, rat membranes exhibit a change in the composition of their lipid components, with polyunsaturated long-chain fatty acids, such as 20:4, 22:5, and 22:6, replacing 18:2 and 18:3; CR has been shown to attenuate this age change (Laganiere & Yu, 1987, 1989, 1993; Merry, 2000; Venkatraman & Fernandes, 1992). Thus CR reduces the susceptibility of membrane lipids to peroxidation, and at the same time maintains membrane fluidity (Pieri, 1997; Yu et al., 1992); the findings of Yu et al. (1992) on mitochondrial membrane fluidity are summarized in Figure 3-5. CR has been found to lessen the age-associated decrease in fluidity of mitochondrial membranes in rat liver (Choe et al., 1995), brain (Choi & Yu, 1995), and heart (Lee et al., 1999). In addition to affecting mitochondrial function, the age-associated decrease in membrane fluidity is likely to influence many other cellular functions, such as the cellular signal transduction systems utilizing plasma membrane receptors. Cavallini et al. (2002) found that CR prevents the age-associated increase in the level of dolichol in the liver of male Sprague–Dawley rats; since dolichol is a lipid that resides in cellular membranes, they suggest that this action of CR has a beneficial effect on the functioning of these membranes. It is clear that further study of CR's effect on cellular membranes, in relation to specific cellular functions, rates a high priority.

Protein structure, function, and turnover

The primary players in an organism's living processes are its many thousands of protein species. The functioning of protein species depends on their three-dimensional structure; and the structures of many, but not all, are modified posttranslationally with increasing age (Gafni, 2001). Modifications are either covalent (oxidation, glycation, glycoxidation,

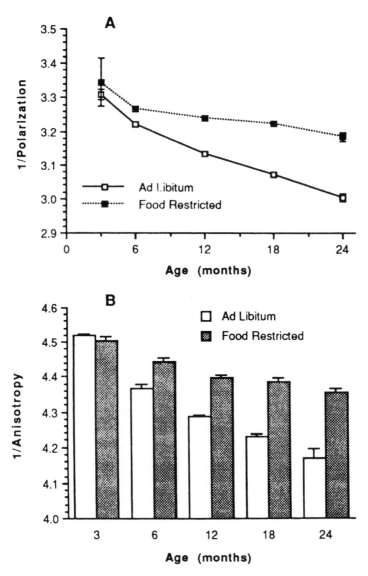

Figure 3-5. The influence of age and CR on mitochondrial membrane fluidity in male F344 rats measured by two methods: (A) 1/polarization and (B) 1/anisotropy. The authors designate long-term CR as food restricted. (From Yu et al., 1992.)

deamidation) or noncovalent (conformation, aggregation); and altered functional characteristics often result from such modifications. It is not known whether these changes in protein structure are the cause or the result of aging.

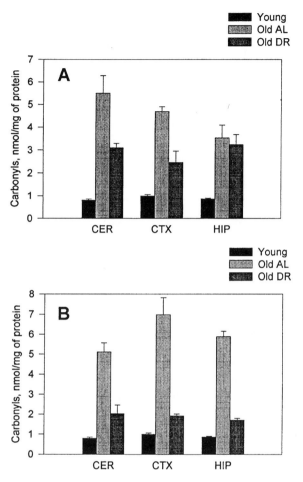

Figure 3-6. Protein carbonyl content of the homogenates of cerebellum (CER), cortex (CTX) and hippocampus (HIP) of male (A) and female (B) BN rats at 3 months and 30 months of age. DR denotes 30-month-old rats that had been on long-term CR. (From Askenova et al., 1998.)

CR is known to decrease the level of oxidatively damaged proteins in mice (Sohal et al., 1994a,b) and rats (Askenova et al., 1998; Youngman et al., 1992). It is particularly effective in preventing oxidative damage of proteins in some regions of mouse brain (striatum, cerebellum, midbrain, and cortex) but not in the hippocampus (Dubey et al., 1996). And in rat brain (Figure 3-6), CR decreases the age-associated increase in the carbonyl content of proteins in the cerebellum and cortex of both genders and in the hippocampus of the females but not the males (Askenova et al., 1998). Lass et al. (1998)

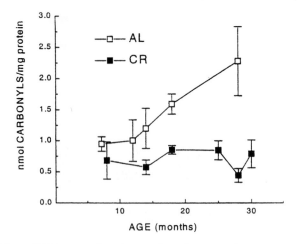

Figure 3-7. Effect of CR started at 4 months of age on the age-associated increase in the carbonyl content of skeletal muscle mitochondria of male C57BL/6 mice. (From Lass et al., 1998.)

reported that CR prevents the age-associated increase in the carbonyl content of mitochondrial proteins in mouse skeletal muscle (Figure 3-7) and it attenuates the increase in *o,o'-dityrosine* in the proteins of mouse cardiac and skeletal muscle (Leeuwenburgh et al., 1997). Forster et al. (2000) found that the carbonyl content increased and the sulfhydryl content of proteins decreased in the heart and brain of mice fed *ad libitum*, and that CR attenuated both of these oxidative changes. However, when mice on the CR regimen in the age range of 15–22 months were then provided food *ad libitum*, the effects of CR on brain protein carbonyl content and heart protein sulfhydryl content completely disappeared within 3–6 weeks, and the effect on the sulfhydryl content of brain proteins partially disappeared. The introduction of a CR regimen for 6 weeks in 15- to 22-month-old *ad libitum*-fed mice reduced the carbonyl content and increased the sulfhydryl content of brain proteins (although not to the levels found with long-term CR) but had no effect on the sulfhydryl content of heart proteins. Similar brief alternations between CR and feeding *ad libitum* did not modify oxidative damage in proteins of mouse skeletal muscle mitochondria (Lass et al., 1998). The protein carbonyl content in the liver of *ad libitum*-fed rats is increased late in life, and CR prevents this increase (Vittorini et al., 1999). Also, 3½ months of CR decreases the carbonyl content of proteins in liver mitochondria of 30-month-old rats (Goto et al., 2002).

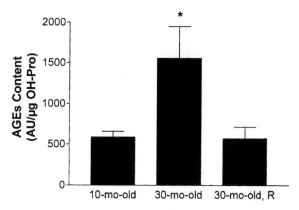

Figure 3-8. The content of advanced glycation end products (AGEs) in liver collagen of 10- and 30-month-old female WAG/Rij rats fed *ad libitum* or on CR starting at 10 months of age. The authors designate CR as R. (From Teillet et al., 2002.)

CR decreases the glycation of hemoglobin in male F344 rats (Masoro et al., 1989) and of skin and aorta collagen in Wistar rats (Misik et al., 1991). It also retards the accumulation of advanced glycation end products in the skin, tail tendon, and aorta collagen of male C57BL/6 mice (Reiser, 1994). Recently, Teillet et al. (2002) reported that CR decreases the age-associated increase in advanced glycation end products in hepatic collagen of female WAG/Rij rats (Figure 3-8). Cefalu et al. (1995) confirmed that CR decreases the glycation of rat hemoglobin, and found that it also decreases the glycation of plasma proteins. They further found that CR decreases the concentration of both N^ε-(carboxymethyl)lysine (CML) and pentosidine (advanced glycation end products) in rat skin collagen. CR inhibits the rate of glycoxidation in rat skin collagen (Sell et al., 1996). Also, the level of Amadori product in the skin of both rats and mice increases rapidly with increasing age, and CR attenuates this increase (Sell, 1997). However, Sell and Monnier (1997) reported that CR had little or no effect on pentosidine formation in the collagen of tail tendon, ear auricle, and skin of male C57BL/6 mice. But CR significantly decreased such formation in the collagen of tail tendon and ear auricle, though not skin, of male DBA/2 mice. Novelli et al. (1998) reported that CR decreased the age-associated accumulation of advanced glycation end products in skin collagen, but not in aorta collagen, of Sprague–Dawley rats. It is apparent that much remains to be done before these scattered findings on protein glycation can be interpreted in a coherent way.

There is no information available on the effects of CR on age-associated noncovalent changes in protein structure. Since such changes play a role in age-associated human diseases (Gafni, 2001), research in this area is much needed.

The accumulation of damaged proteins with age could be due to the decrease in protein degradation (proteolysis) with increasing age (Van Remmen et al., 1995). CR enhances proteolysis, and this action may well play a major role in decreasing the level of damaged proteins (Ward, 1988). The proteosome, a multienzymatic proteolytic complex, is known to play a key role in the degradation of damaged proteins (Friguet et al., 2000). While it, therefore, seems likely that CR enhances proteolysis by modulating the proteosome, it is difficult to interpret the results of studies designed to test this hypothesis (Scrofano et al., 1998; Shibatani et al., 1996). However, CR has been found to prevent the age-associated increase in ChT-L and T-L activities in rat liver, and to maintain a higher PGPH activity throughout life (Shibatani et al., 1996). Also, CR increases the mRNA levels coding for the proteosomal subunit TBP1 and the proteosomal activator subunit PA28 in mouse skeletal muscle (Lee et al., 1999). Recently, Goto et al. (2002) reported that CR of 3½ months duration in 30-month-old rats increases the activity of hepatic proteosomes.

Autophagy is a process that sequesters and degrades damaged organelles and macromolecules (Seglen et al. 1990). CR increases autophagy in rat liver and attenuates its age-associated decline (Cavallini, et al., 2001; Donati et al., 2001).

Turnover of body protein is increased by CR in rats and mice, and this increase is sustained until old ages (Dhahbi et al., 2001; Lewis et al., 1985). Indeed, CR of a few months duration increases protein turnover in old rats (Goto et al., 2002). When old rats are maintained on long-term CR, protein synthesis by liver and kidney cells is greater than that of rats of the same age on an *ad libitum* dietary regime (Birchenall-Sparks et al., 1985; Ricketts et al., 1985). The increased protein turnover may contribute to CR's anti-aging action by decreasing accumulation of damaged proteins (Weindruch et al., 2001).

Cell proliferation

A decrease in the number of cells of a particular type (e.g., muscle fibers) underlies some of the functional deterioration of the aging phenotype (Wolf & Pendergrass, 1999). Moreover, the age-associated increase in the number of cells (e.g., hyperplasia and neoplasia) also results in dysfunction (Clark, 1999). Obviously, it is vital for optimal function that the generation of new cells (cell proliferation) and the death of existing cells be in balance.

In mice, the proliferation of liver hepatocytes, kidney tubule cells, and pancreas acinar cells decreases with increasing age (Wolf et al., 1995). CR decreases the proliferation of many cell types in young mice (Lok et al., 1990). However, when old mice on a long-term CR regimen receive food *ad libitum* for 4 weeks, proliferation of hepatocytes, kidney tubule cells, and pancreas acinar cells increases to levels near those seen in young *ad libitum*-fed mice (Wolf et al., 1995). Thus, CR maintains the youthful potential for cell proliferation, a potential expressed when an immediate source of additional dietary energy becomes available. Furthermore, in some cell types (e.g., bone marrow stromal endothelial-like cells and bone marrow osteoblasts), the additional energy source is not needed to manifest the enhancing effect of CR on cell proliferation in old mice. CR also preserves the proliferation capacity of mouse lens epithelial cells (Li et al., 1997). However, in contrast to the findings with mice, CR does not preserve the proliferation potential in aging rhesus monkeys (Pendergras et al., 1999).

In vitro studies of clonal cellular replication capacity were carried out using mouse tail skin fibroblasts, kidney epithelial cells, spleen fibroblasts, bone marrow fibroblasts, and bone marrow endothelium (Pendergrass et al., 1995). After 14 days in culture, the clone size reached by each of these cell types was found to decrease with the age of the donor mouse. At advanced ages, however, the clone sizes of cells from mice on CR were found to be much larger than those from mice of the same age fed *ad libitum*. In contrast, Pignolo et al. (1992) found that CR does not prevent the age-associated decrease in proliferation of rat skin fibroblasts when studied in mass culture.

The studies discussed above suggest that CR acts to maintain a youthful potential for cell proliferation for most of the mouse life span. This action may serve to prevent or retard the occurrence of physiological and pathological conditions that stem from a reduction in the number of a particular cell type. However, aging is associated with an increase in epithelial cell proliferation in rat small intestine (Holt & Yeh, 1989) and colon (Holt & Yeh, 1988). By delaying this age-associated change (Heller et al., 1990), CR may protect against an age-associated increase in intestinal and colon cancer. A study by Lu et al. (2002) of 28-month-old male F344 rats has shown that the effects of 14 weeks of CR on cellular proliferation in the organs of the digestive system are complex; they found that CR in rats inhibits cellular proliferation in the glandular stomach

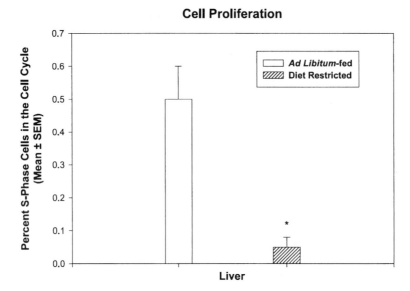

Figure 3-9. The effects of 14 weeks of CR on cell proliferation in the liver of male 28-month-old F344 rats. The authors designate 14 weeks of CR as diet restricted. (From Lu et al., 2002.)

and liver, but enhances it in the mucosal tissue of the duodenum and forestomach (Figures 3-9 and 3-10). Also, rats on a long-term CR regimen maintain a smaller number of adipocytes in their fat depots throughout life (Masoro, 1992).

Apoptosis

The opposite side of the coin of cell proliferation is cell death. It involves two distinct processes: necrosis and apoptosis (Warner, 1999). Necrosis refers to the massive cell death caused by severe injury, such as that following a stroke or coronary thrombosis. Apoptosis refers to genetically programmed cell death, which serves to carry out important physiological functions, such as cell number homeostasis and removal of damaged cells. In regard to aging, excessive apoptosis can cause an undesirable decrease in the number of a particular cell type, such as the age-associated loss of neurons in particular regions of the central nervous system. On the other hand, too little apoptosis can lead to an accumulation of damaged cells, hyperplasia, and/or neoplasia. Warner et al. (1995) hypothesized that CR's anti-aging action could be due to the upregulation of apoptosis.

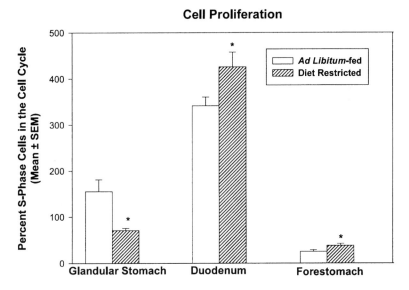

Figure 3-10. The effects of 14 weeks of CR on cell proliferation in the glandular stomach, duodenum, and forestomach of male 28-month-old F344 rats. The authors designate 14 weeks of CR as diet restricted. (From Lu et al., 2002.)

Table 3-3
Frequencies of apoptotic bodies (AB) in livers of 12-month-old mice

Diet group	%AB
Ad libitum-fed	0.023
CR	0.008

Note: Data from James and Muskhelishvilli, 1994.

CR has been found to promote apoptosis in the liver of aging mice (James & Muskhelishvilli, 1994; Muskhelishvilli et al., 1995) and small intestine and colon of aging rats (Holt et al., 1998). The findings of James and Muskhelishvilli (1994) are summarized in Table 3-3. Grasyl-Kraup et al. (1994) reported that CR preferentially enhances the apoptosis of preneoplastic cells in the liver of rats. However, Higami et al. (1997) found that CR somewhat reduces the level of apoptosis in hepatocytes of aging rats. Also, in response to the administration of cycloheximide, CR was found to prevent the age-associated increase in the susceptibility of rat hepatocytes to apoptosis.

As mentioned above, Warner et al. (1995) hypothesized that CR's anti-aging action could be due to the upregulation of apoptosis. James et al. (1998) second this view. Recently, Zhang and Herman (2002) reviewed the evidence in support of this hypothesis and concluded that the currently available data are not sufficient to draw a conclusion on its validity. Clearly, the influence of CR on this important cellular process warrants further study.

Cellular signal transduction

Hormones, neurotransmitters, and cytokines regulate the activities of cells, and they are involved in both the coordinated functioning of multicellular organisms and the responses of such organisms to the environment. These humoral agents interact with cells by binding to specific receptors, some of which are located on the cell surface and others within the cell. This binding brings into play a complex pathway of interlinked chemical reactions, termed a cellular signaling pathway. It is through such pathways that humoral agents influence the activities of cells. With increasing age, there are changes in receptors and/or other components of signaling pathways.

The influence of age and CR on cellular signal transduction in the central nervous system has been the subject of several studies. Roth et al. (1984) reported there is an age-associated decrease in dopamine receptors in the corpus striatum of the rat, and that CR retards the loss of these receptors. On the other hand, May et al. (1992) found that while a decrease in striatal dopamine receptors also occurs in mice with age, CR does not attenuate the loss in this species. In regard to the cholinergic system, CR increases the level of muscarinic binding sites in the striata of old rats (London et al., 1985).

The phosphatidylinositol-4,5-biphosphate-phospholipase C-inositol-1,4,5-triphosphate system is known to play a key role in cellular signaling. Using slices of brain cortex and corpus striatum from young and old rats, Undie and Friedman (1993) found a decreased ability of a cholinergic agonist and a dopaminergic agonist to promote phosphoinositide hydrolysis at advanced age, and that CR attenuates this age-associated change.

Mitogen-activated protein kinases (MAPKs) also play a central role in cellular signaling. Three MAPK subfamilies have been identified: extracellular signal-regulated kinases (ERK), p38 MAPK, and c-Jun *N*-terminal kinases (JNK). Zhen et al. (1999) found that basal activities of ERK and p38 MAPK are decreased in the brain of old rats, and that the activation of ERK in response to stimuli is reduced in cortical brain slices of old rats. CR prevents the age-associated decline in basal ERK phosphorylation and kinase activity, and it attenuates the reduction in p38 MAPK activity. Thus it is clear that CR modulates changes with age in the signaling pathways of the nervous system. While the exploration of this subject is still only beginning, further study holds promise of providing fundamental information on brain aging and its modulation by CR.

CR has also been found to modulate cellular signaling in the growth hormone-IGF-1 axis of rodents (Sonntag et al., 1999). In mice, there is progressive impairment with age in the growth hormone receptor signal transduction pathway, and CR delays this impairment (Xu & Sonntag, 1996a). CR blunts the age-associated decrease in growth hormone receptor phosphorylation. It also decreases both growth hormone-induced activation and nuclear translocation of Stat-3 (Xu & Sonntag, 1996b). In addition, CR increases the density of type 1 IGF-1 receptors in liver, heart, and skeletal muscle (D'Costa et al., 1993).

Indeed, CR has been found to modulate age-associated changes in cellular signaling in several other tissues. With increasing age, there is a decreased ability of epinephrine to increase the cytoplasmic concentration of calcium ion in rat parotid gland acinar cells, and CR does not prevent this age change, though it does slow it (Salih et al., 1997). The ability of epinephrine to elevate the level of c-AMP in rat smooth muscle cells in culture decreases with the increasing age of the donor animal; the extent of this age change is less in cells from rats on long-term CR (Volicer et al., 1983). Scarpace and Yu (1987) found that CR retards the age-associated loss of beta-adrenergic receptors in rat lung; it also prevents the age-associated decrease in isoproterenol- and epinephrine-stimulated adenylate cyclase activity in this organ. With age, the activation by concanavalin A of two important components of cellular signaling pathways, MAPK and calcineurin, decreases in rat splenic T cells, and this is partially prevented by CR (Pahlavani & Vargas, 2000).

It appears certain that CR influences age-changes in cellular signaling in a variety of cell types. As to the future, knowledge of the bewildering complex subject of cellular signaling is expanding incredibly rapidly making it likely that further research on this subject may soon provide important insights on the cellular and molecular basis of CR's anti-aging action.

References

Aidoo, A., Mittelstaedt, R. A., Lyn-Cook, L. E., Duffy, P. H., & Heflich, R. H. (1999). Effect of dietary restriction on lymphocyte Hprt mutant frequency in aging rats. *J. Environ. Mol. Mutagenetics* 33: 5.

Aly, K. B., Pipkin, J. L., Hinson, W. G., Feuers, R. J., Duffy, P. H., Lyn-Cook, L. E., & Hart, R. W. (1994). Chronic caloric restriction induces stress proteins in the hypothalamus of rats. *Mech. Ageing Dev.* 76: 11–23.

Askenova, M. V., Askenov, M. Y., Carney, J. M., & Rutterfield, D. A. (1998). Protein oxidation and enzyme activity decline in old Brown Norway rats are reduced by dietary restriction. *Mech. Ageing Dev.* 100: 157–168.

Aspnes, L. E., Lee, C. M., Weindruch, R., Chung, S. S., Roecker, E. B., & Aiken, J. M. (1997). Caloric restriction reduces fiber loss and mitochondrial abnormalities in aged rat muscle. *FASEB J.* 11: 573–581.

Beckman, K. B. & Ames, B. N. (1998). Mitochondrial aging: Open questions. *Ann. N. Y. Acad. Sci.* 854: 118–127.

Birchenall-Sparks, M. C., Roberts, M. S., Staecker, R. J., Hardwick, J. P., & Richardson, A. (1985). Effect of dietary restriction on liver protein synthesis in rats. *J. Nutrition 115*: 944–950.

Bohr, V. A. & Anson, R. M. (1995). DNA damage, mutation and fine structure DNA repair in aging. *Mutation Res.* 338: 25–34.

Cao, S. X., Dhahbi, J. M., Mote, P. L., & Spindler, S. R (2001). Genomic profiling of short–and long-term caloric restriction effects in liver of aging mice. *Proc. Natl. Acad. Sci. USA 98*: 10630–10635.

Casciano, D. A., Chou, M., Lyn-Cook, L. E. & Aidoo, A. (1996). Calorie restriction modulates chemically induced *in vivo* somatic mutation frequency. *Environ. Mol. Mutagen.* 27: 162–164.

Cavallini, G., Donati, A., Gori, Z., Parentini, I., & Bergamini, E. (2002). Low level dietary restriction retards age-related dolichol accumulation. *Aging Clin. Exp. Res. 14*: 152–154.

Cavallini, G., Donati, A., Gori, Z., Pollera, M. & Bergamini, E. (2001). The protection of rat liver autophagic proteolysis from the age-related decline co-varies with the duration of the anti-ageing food restriction. *Exp. Gerontol.* 36: 497–506.

Cefalu, W. T., Bell-Farrow, A. D., Wang, Z. Q., Sonntag, W. E., Fu, M-X., Baynes, J. W., & Thorpe, S. R. (1995). Caloric restriction decreases age-dependent accumulation of glycoxidation products, N^ε-(Carboxymethyl)lysine and pentosidine, in rat skin collagen. *J. Gerontol.: Biol. Sci. 50A*: B337–B341.

Chatterjee, B., Fernandes, G., Yu, B. P., Song, C., Kim, J. H., Demyan, W., & Roy, A. K. (1989). Calorie restriction delays age-dependent loss in androgen responsiveness of rat liver, *FASEB J. 3*: 169–173.

Choe, M., Jackson, C., & Yu, B. P. (1995). Lipid peroxidation contributes to age-related membrane rigidity. *Free Radic. Biol. Med.* 18: 977–984.

Choi, J. H. & Yu, B.P. (1995). Brain synaptosomal aging: free radicals and membrane fluidity. *Free Radic. Biol. Med.* 18: 133–139.

Chung, M. H., Kasai, H., Nishimura, S., & Yu, B. P. (1992). Protection of DNA damage by dietary restriction. *Free Radic. Biol. Med.* 12: 523–525.

Clark, W. R. (1999). *A Means to an End. The Biological Basis of Aging and Death*. New York: Oxford University Press.

D'Costa, A. P., Lenham, J. E., Ingram, R. L., & Sonntag, W. E. (1993). Moderate caloric restriction increases type 1 IGF receptors and protein synthesis in aging rats. *Mech. Ageing Dev. 71*: 59–71.

Dempsey, H., Pfeiffer, M, & Morley, A. A. (1993). Effect of dietary restriction on *in vivo* somatic mutation in mice. *Mutation Res.* 291: 141–145.

Desai, V. G., Weindruch, R., Hart, R. W., & Feuers, R. J. (1996). Influences of age and dietary restriction on the gastrocnemius electron transport system activities in mice. *Arch. Biochem. Biophys. 333*: 144–151.

Dhahbi, J. M., Mote, P. L., Wingo, J., Rowley, B. C., Cao, S., Walford, R. L., & Spindler, S. R. (2001). Caloric restriction alters the feeding response of key metabolic enzyme genes. *Mech. Ageing Dev. 122*: 1033–1048.

Dhahbi, J. M., Mote, P. L. Wingo, J., Tillman, J. B., Walford, R. L., & Spindler, S. R. (1999). Calories and aging alter gene expression for gluconeogenic, glycolytic, and nitrogen-metabolizing enzymes. *Am. J. Physiol. 277*: E352–E360.

Djuric, Z., Lu, M. H., Lewis, S. M., Luongo, D. A., Chen, X. W., Heilbrun, L. A., Reading, B. A., Duffy, P. H., & Hart, R. W. (1992). Oxidative damage levels in rats fed low-fat, high fat, or caloric restricted diets. *Toxicol. Appl. Pharmacol. 115*: 156–160.

Donati, A., Cavallini, G., Paradiso, C., Vittorini, J.S., Pollera, M., Gori, Z., & Bergamini, E. (2001). Age-related changes in autophagic proteolysis of rat isolated liver cells: Effects of antiaging dietary restrictions. *J. Gerontol.: Biol. Sci. 56A*: B375–B383.

Dubey, A., Forster, M. J., Lal, H., & Sohal, R. S. (1996). Effect of age and caloric intake on protein oxidation in different brain regions and on behavioral functions of the mouse. *Arch. Biochem. Biophys. 333*: 189–197.

Failla, G. (1958). The aging process and cancerogenesis. *Ann. N. Y. Acad. Sci. 71*: 1124–1135.

Feuers, R. J. (1998). The effects of dietary restriction on mitochondrial dysfunction in aging. *Ann. N.Y Acad. Sci. 854*: 192–201.

Forster, M. J., Sohal, B. H., & Sohal, R. S. (2000). Reversible effects of long-term caloric restriction on protein oxidative damage. *J. Gerontol.: Biol. Sci. 55A*: B522–B529.

Friedberg, E. C. (1996). Relationships between DNA repair and transcription. *Ann. Rev. Biochem. 65*: 15–42.

Friguet, B., Bulteau, A-L., Chondrogianni, N., Conconi, M., Petropoulos, I. (2000). Protein degradation by the proteosome and its implications in aging. *Ann. N.Y Acad. Sci. 908*: 143–154.

Gafni, A. (2001). Protein structure and turnover. In: E. J. Masoro & S. N. Austad (Eds.), *Handbook of the Biology of Aging*, 5th ed. (pp. 59–83). San Diego: Academic Press.

Goto, S., Takahashi, R., Araki, S., & Nakamoto, H. (2002). Dietary restriction initated in late adulthood can reverse age-related alterations of protein and protein metabolism. *Ann. NY Acad. Sci. 959*: 50–56.

Grasyl-Kraup, B., Bursch, W., Ruttky-Nedecky, B., Wagner, A., Lauer, B. C., & Schulte-Hermann, R. (1994). Food restriction eliminates preneoplastic cells though apoptosis and antagonizes carcinogenseis in rat liver. *Proc. Natl. Acad. Sci. U.S.A 91*: 9995–9999.

Guo, Z. M., Heydari, A., Richardson, A. (1998a). Nucleotide excision repair of actively transcribed versus nontranscribed DNA in rat hepatocytes: Effect of age and dietary restriction. *Exp. Cell Res. 245*: 228–238.

Guo, Z. M., Van Remmen, H., Wu, W-T., & Richardson, A. (1998b). Effect of cAMP-induced transcription on the repair of the phosphoenolpyruvate carboxykinase gene by hepatocytes isolated from young and old rats. *Mutation Res. 409*: 37–48.

Haley-Zitlin, V. & Richardson, A. (1993). Effect of dietary restriction on DNA repair and DNA damage. *Mutation Res. 295*: 237–245.

Han, E-S., Hilsenbeck, S. G., Richardson, A., & Nelson, J. F. (2000). CDNA expression arrays reveal incomplete reversal of age-related changes in gene expression by caloric restriction. *Mech. Ageing Dev. 115*: 157–174.

Heller, T. D., Holt, P. R., & Richardson, A. (1990). Food restriction retards age-related histological changes in rat small intestine. *Gastroenterology 98*: 387–391.

Heydari, A. R., Conrad, C. C. & Richardson, A. (1995). Expression of heat shock genes in hepatocytes is affected by age and food restriction, *J. Nutrition. 125*: 410–418.

Heydari, A. R., Wu, B., Takahashi, R., Strong, R. & Richardson, A. (1993). Expression of heat shock protein 70 is altered by age and diet at the level of transcription. *Mol. Cell. Biol. 13*: 2909–2918.

Heydari, A. R., You, S., Takahashi, R., Gutsmann-Conrad, A., Sarge, K. D., & Richardson, A. (2000). Age-related alterations in the activation of heat shock transcription factor 1 in rat hepatocytes. *Exp. Cell Res. 256*: 83–93.

Higami, Y., Shimokawa, I., Okimoto, T., Tomita, M., Yuo, T. & Ikeda, T. (1996). Susceptibility of hepatocytes to cell death by single administration of cycloheximide in young and old F344 rats. Effect of dietary restriction. *Mutation. Res. 357*: 225–230.

Higami, Y., Shimokawa, I., Okimoto, T., Tomita, M., Yuo, T., & Ikeda, T. (1997). Effect of aging and dietary restriction on hepatocyte proliferation and death in male F344 rats. *Cell Tissue Res. 288*: 69–77.

Holt, P. R. & Yeh, K. Y. (1988). Colonic proliferation is increased in senescent rats. *Gastroenterology 95*: 1556–1563.

Holt, P. R. & Yeh, K. Y. (1989). Small intestinal crypt cell proliferation rates are increased in senescent rats. *J. Gerontol.: Biol. Sci. 44*: B9–B14.

Holt, P. R., Moss, S. F., Heydari, A. R., & Richardson, A. (1998). Diet restriction increases apoptosis in the gut of aging rats. *J. Gerontol.: Biol. Sci. 53A*: B168–B172.

James, S. J. & Muskhelishvilli, L. (1994). Rates of apoptosis and proliferation vary with caloric intake and may influence incidence of spontaneous hepatoma in C57BL/6xC3HF1 mice. *Cancer Res. 54*: 5508–5510.

James, S. J., Muskhelishvilli, L, Gaylor, D. W., Turturro, A. & Hart, R. (1998). Upregulation of apoptosis with dietary restriction: implications for carcinogenesis and aging. *Environ. Health Persp. 106*: 307–312.

Kaneko, T., Tahara, S., & Matsuo, M. (1997). Retarding effect of dietary restriction on the accumulation of 8-hydroxy-2′-deoxyguanosine in organs of Fischer 344 rats during aging. *Free Radic. Biol. Med. 23*: 76–81.

Kang, C-M., Kristal, B. S., & Yu, B. P. (1998). Age-related mitochondrial DNA deletions: Effect of dietary restriction. *Free Radic. Biol. Med. 24*: 148–154.

Kayo, T., Allison, D. B. Weindruch, R. & Prolla, T. A. (2001). Influences of aging and caloric restriction on the transcriptional profile of skeletal muscle from rhesus monkeys. *Proc. Natl. Acad. Sci. U.S.A 98*: 5093–5098.

Kristal, B. S. & Yu, B. P. (1998). Dietary restriction augments protection against the induction of mitochondrial permeability transition. *Free Radic. Biol. Med. 24*: 1269–1277.

Laganiere, S. & Yu, B. P. (1987). Anti-lipoperoxidation action of food restriction. *Biochem. Biophys. Res. Commun. 145*: 1185–1191.

Laganiere, S. & Yu, B. P. (1989). Effects of chronic food restriction in aging rats. I. Liver subcellular membranes. *Mech. Ageing Dev. 48*: 207–219.

Laganiere, S. & Yu, B. P. (1993). Modulation of membrane phospholipid fatty acid composition by age and food restriction. *Gerontology 39*:7–18.

Lal, S. B., Ramsey, J. J., Momemdjou, S, Weindruch, R. & Harper, M-E. (2001). Effect of caloric restriction on skeletal muscle mitochondrial proton leak in aging rats. *J. Gerontol.: Biol. Sci. 56A*: B116–B122.

Lass, A., Sohal, B. H., Weindruch, R., Forster, M. J., & Sohal, R. S. (1998). Caloric restriction prevents age-associated accrual of oxidative damage to mouse skeletal muscle mitochondria. *Free Radic. Biol. Med. 25*: 1089–1097.

Lee C-K., Klopp, R. G., Weindruch, R., & Prolla, T.A (1999). Gene expression profile of aging and its retardation by caloric restriction. *Science 285*: 1390–1393.

Lee C-K., Weindruch, R. & Prolla, T. A. (2000). Gene-expression profile of the ageing brain in mice. *Nature Genet. 25*: 294–297.

Lee, J., Yu, B. P., & Herlihy, J. T. (1999). Modulation of cardiac mitochondrial fluidity by age and calorie intake. *Free Radic. Biol. Med. 26*: 260–265.

Leeuwenburgh, C., Wagner, P., Holloszy, J. O., Sohal, R. S., & Heinecke, J. W. (1997). Caloric restriction attenuates cross-linking of cardiac and skeletal muscle proteins in aging mice. *Arch. Biochem. Biophys. 346*: 74–80.

Lewis, S. E., Goldspink, D. F., Phillips, J. G., Merry, B. J., & Holehan, A. M. (1985). The effects of aging and chronic dietary restriction on whole body growth and protein turnover in the rat. *Exp. Gerontol. 20*: 253–263.

Li, Y., You, Q. R., & Wolf, N, S. (1997). Long-term caloric restriction delays age-related decline in proliferation capacity of mouse lens epithelial cells *in vitro* and *in vivo*. *Invest. Ophthal. & Visual Sci. 38*: 100–107.

Licastro, F., Weindruch, R., Davis, L. J., & Walford, R. L. (1988). Effect of dietary restriction upon age-associated decline of lymphocyte DNA repair activity in mice. *Age 11*: 48–52.

Lipman, J. M., Turturro, A., & Hart, R. W. (1989). The influence of dietary restriction on DNA repair in rodents: a preliminary study. *Mech. Ageing Dev. 48*: 135–143.

Lok, F., Scott, F. W., Margeau, R., Malcolm, S. & Clayson, D. B. (1990). Caloric restriction and cellular proliferation in various tissues of the female Swiss Webster mouse. *Cancer Letts. 51*: 67–73.

London, E. D., Waller, S. B., Ellis, A. T., & Ingram, D. K. (1985). Effects of intermittent feeding on neurochemical markers of aging rat brain. *Neurobiol. Aging 6*: 194–204.

Lu, M. H., Warbritton, A., Tang, N. & Bucci, T. G. (2002). Dietary restriction alters cell proliferation in rats: an immunohistochemical study of labeling proliferating cell nuclear antigen. *Mech. Ageing Dev. 123*: 391–400.

Masoro, E. J. (1992). Aging and proliferation homeostasis: modulation by food restriction in rodents. *Lab. Animal Sci. 42*: 132–137.

Masoro, E. J., Katz, M. S., & McMahan, C. A. (1989). Evidence for the glycation hypothesis of aging from the food-restricted rodent model. *J. Gerontol.: Biol. Sci. 44*: B20–B22.

May, P. C., Telford, N., Salo, D., Anderson, C., Kohama, S. G., Finch, C., Walford, R. A., & Weindruch, R. (1992). Failure of dietary restriction to retard age-related neurochemical changes in mice. *Neurobiol. Aging 13*: 787–791.

Merry, B. J. (2000). Caloric restriction and age-related oxidative stress. *Ann. N.Y Acad. Sci. 908*: 180–198.

Misik, I., Struzinsky, R., & Deyl, Z. (1991). Change with age of UV absorbance and accumulation of ε–hexosyllysine in collagen from Wistar rats living on different food restriction regimes. *Mech. Aging Dev. 57*: 163–174.

Moore, S. A., Lopez, A., Richardson, A., & Pahlavani, M. A. (1998). Effect of age and dietary restriction on expression of heat shock protein 70 in rat alveolar macrophages. *Mech. Ageing Dev. 104*: 59–73.

Muskhelishvilli, L., Hart, R. W., Turturro, A. & James, S. J. (1995). Age-related changes in the intrinsic rate of apoptosis in livers of diet-restricted and *ad libitum*-fed B6C3F1 mice. *Am. J. Path. 147*: 20–24.

Naeim, F. & Walford, R. L. (1985). Aging and cell membrane complexes: The lipid bilayer, and integral proteins, and cytoskeleton. In: C. E. Finch & E. L. Schneider (Eds.), *Handbook of the Biology of Aging*, 2nd ed. (pp. 272–289). New York: Van Nostrand Reinhold.

Novelli, M., Masiello, P., Bombara, M., & Bergamini, E. (1998). Protein glycation in aging male Sprague-Dawley rat: Effects of antiaging diet restrictions. *J. Gerontol. Biol. Sci. 53A*: B94–B100.

Olgun, A., Akman, S., Serdar, M. A., & Kutluay, T. (2002). Oxidative phosphorylation enzyme complexes in caloric restriction. *Exp. Gerontol. 37*: 639–645.

Pahlavani, M. A. & Vargas (2000). Influences of aging and caloric restriction on activation of Ras/MAPK, Calcineurin, and CaMK-IV activities in rat T cells. *Proc. Soc. Exp. Biol. 223*: 163–169.

Pahlavani, M. A., Harris, M. D., & Richardson, A. (1996). Expression of heat shock protein 70 in rat spleen lymphocytes is affected by age but not by food restriction. *J. Nutrition 126*: 2069–2075.

Pahlavani, M. A., Harris, M. D., & Richardson, A. (1997). The increase in the induction of IL-2 expression with caloric restriction is correlated to changes in the transcription factor NFAT. *Cell. Immun. 180*: 10–19.

Pendergrass, W. R., Lane, M. A., Bodkin, N. L., Hansen, B. C., Ingram, D. K., Roth, G.S., Yi, L., Bin, H., & Wolf, N. S. (1999). Cellular proliferation potential during aging and caloric restriction in rhesus monkey *(Macaca mulatta)*. *J. Cell. Physiol. 180*: 123–130.

Pendergrass, W. R., Li, Y., Jiang, D., Fei, R. G., & Wolf, N. S. (1995). Caloric restriction: conservation of cellular replicative capacity *in vitro* accompanies life extension in mice. *Exp. Cell Res. 217*: 309–316.

Pieri, C. (1997). Membrane and lipid peroxidation in food restricted animals. *Age 20*: 71–79.

Pieri, C., Recchioni, R., Moroni, F., Marcheselli, F., Flasca, M., & Piantanelli, L. (1990). Food restriction in female Wistar rats I. Survival characteristics, membrane micro-viscosity, and proliferative response in lymphocytes. *Arch. Gerontol. Geriatr. 11*: 99–108.

Pignolo, R. J., Masoro, E. J., Nichols, W. W., Bradt, C. I., & Cristofalo, V. J. (1992). Skin fibroblasts from Fischer 344 rats undergo similar changes in replicative life span but not immortalization with caloric restriction of donors. *Exp. Cell Res. 201*: 16–22.

Ramsay, G. (1998). DNA chips: state-of-the art. *Nature Genetics 16*: 40–44.

Rao, G., Xia, E., Nadakanukaren, M. J., & Richardson, A. (1990). Effect of dietary restriction on age-dependent changes in the expression of antioxidant enzymes in rat liver. *J. Nutrition 120*: 602–609.

Rao, K. S., Ayyagari, S., Raji, N. S., & Murthy, K. J. R. (1996). Undernutrition and aging: Effects on DNA repair in human peripheral lymphocytes. *Curr. Sci. 71*: 464–469.

Reiser, K. M. (1994). Influence of age and long-term dietary restriction on enzymatically mediated crosslinks and nonenzymatic glycation of collagen in mice. *J. Gerontol.: Biol. Sci. 49*: B71–B79.

Richardson, A., Butler, J. A., Rutherford, J. S., Semsei, I., Gu, M. Z., Fernandes, G., & Chiang, W. H. (1987). Effect of age and dietary restriction on the expression of α_{2v}-globulin. *J. Biol. Chem. 262*: 12821–12825.

Ricketts, W. G., Birchenall-Sparks, M. C., Hardwick, J. P., & Richardson, A. (1985). Effect of age and dietary restriction on protein synthesis by isolated kidney cells. *J. Cell. Physiol. 125*: 492–498.

Roth, G. S., Ingram, D. K., & Joseph, J. A. (1984). Delayed loss of striatal dopamine receptors during aging of dietarily restricted rats. *Brain Res. 300*: 27–32.

Salih, M. A., Herbert, D. D., & Kalu, D. N. (1993). Evaluation of the molecular and cellular basis for modulation of thyroid C-cell hormone by aging and food restriction. *Mech. Ageing Dev. 70*: 1–21.

Salih, M. A., Kalu, D. N., & Smith, T. C. (1997). Effects of age and food restriction on calcium signaling in parotid acinar cells of Fischer 344 rats. *Aging Clin. Exper. Res. 9*: 419–427.

Scarpace, P. J. & Yu, B. P. (1987). Diet restriction retards the age-related loss of beta-adrenergic receptors and adenylate cyclase activity in rat lung. *J. Gerontol. 42*: 442–446.

Scrofano, M., Shang, F., Nowell, Jr. T. R., Gong, X., Smith, D. E., Kelliher, M., Dunning, J., Mura, C. V., & Taylor, A. (1998). Aging, calorie restriction and ubiquitin-dependent proteolysis in livers of Emory mice. *Mech. Ageing Dev. 101*: 277–296.

Seglen, P, O., Gordon, P. B., & Holen, I. (1990). Non-selective autophagy. *Semin. Cell Biol. 1*: 441–448.

Sell, D. R. (1997). Ageing promotes the increase of early glycation Amadori product as assessed by ε–N-(2-furoylmethyl)-L-lysine (furosine) levels in rodent skin collagen, The relationship to dietary restriction and glycoxidation. *Mech. Ageing Dev. 95*: 81–99.

Sell, D. R. & Monnier (1997). Age-related association of tail tendon break time with tissue pentosidine in DBA/2 vs. C57BL/6 mice: The effect of dietary restriction. *J. Gerontol.: Biol. Sci. 52A*: B277–B284.

Sell, D. R., Lane, M. A., Johnson, W. A., Masoro, E. J., Mock, O. B., Reiser, K. M., Fogarty, J. F., Cutler, R. G., Ingram, D. K., Roth, G. S., & Monnier, V. M. (1996). Longevity and the genetic determination of collagen glycoxidation kinetics in mammalian senescence. *Proc. Natl. Acad. Sci., USA 93*: 485–490.

Shibatani, T., Nazir, M. & Ward, W. F. (1996). Alteration of rat liver 20S proteosome activities by age and food restriction. *J. Gerontol.: Biol. Sci. 51A*: B316–B322.

Sohal, R. S. & Dubey, A. (1994). Mitochondrial oxidative damage, hydrogen peroxide release, and aging. *Free Radic. Biol. Med. 16*: 621–626.

Sohal, R. S., Agarwal, S., Candas, M., Forster, M. J., & Lal, H. (1994a). Effect of age and caloric restriction on DNA oxidative damage in different tissues of C57BL/6 mice. *Mech. Ageing Dev. 76*: 215–224.

Sohal, R. S., Ku, H-H, Agarwal, S., Forster, M. J., & Lal, H. (1994b). Oxidative damage, mitochondrial oxidant generation, and antioxidant defenses during aging and in response to food restriction in the mouse. *Mech. Ageing Dev. 74*: 121–133.

Sonntag, W. E., Lynch, C. D., Cefalu, W. T., Ingram, R. L., Bennett, S. A., Thornton, P. L., & Khan, A. S. (1999). Pleiotropic effects of growth hormone and insulin-like growth factor-I on biological aging: Inferences from moderate caloric-restricted animals. *J. Gerontol.: Biol. Sci. 54A*: B521–B538.

Spindler, S. R., Crew, M. D., Mote, P. L., Grizzle, J. M., & Walford, R. L., (1990). Dietary energy restriction in mice reduces hepatic expression of glucose-regulated protein 78 (BiP) and 94 mRNA. *J. Nutrition. 129*: 1412–1417.

Srivastava, V. K. & Busbee, D. (1992). Decreased fidelity of DNA polymerases and decreased DNA excision repair in aging mice: Effects of caloric restriction. *Biochem. Biophys. Res. Commun. 182*: 712–721.

Szilard, L. (1959). On the nature of aging process. *Proc. Natl. Acad. Sci. USA 45*: 30–45.

Teillet, L., Ribiere, P, Gouraud, S., Bakala, H. & Corman, B. (2002). Cellular signaling, AGE accumulation, and gene expression in hepatocytes of lean aging rats fed ad libitum or food restricted. *Mech. Ageing Dev. 123*: 427–439.

Tilley, R., Miller, S. Srivastava, V. K., & Busby, D. (1992). Enhanced unscheduled DNA synthesis by secondary cultures of lung cells established from calorically restricted aged rats. *Mech. Ageing Dev. 63*: 165–176.

Tillman, J. B., Mote, P. L., Dhahbi, J. M., Walford, R. L., & Spindler, S. R. (1996). Dietary energy restriction in mice negatively regulates hepatic glucose-regulated protein 78 (GRP78) expression at the posttranscriptional level. *J. Nutrition 129*: 416–423.

Undie, A. S. & Friedman, E. (1993). Diet restriction prevents aging-induced deficits in brain phosphoinositide metabolism. *J. Gerontol.: Biol. Sci. 48*: B62–B67.

Van Remmen, H. & Ward, W. F. (1998). Effect of dietary restriction on hepatic and renal phosphenolpyruvate carboxykinase induction in young and old Fischer 344 rats. *Mech. Ageing Dev. 104*: 263–275.

Van Remmen, H., Ward, W. F., Sabia, R. V., & Richardson, A. (1995). Gene expression and protein degradation, In: E. J. Masoro (Ed.), *Handbook of Physiology*, Section 11, *Aging* (pp. 171–234). New York: Oxford University Press.

Venkatramen, J. & Fernandes, G. (1992). Modulation of age-related alterations in membrane composition and receptor-associated immune functions by food restriction in Fischer 344 rats. *Mech. Ageing Dev. 63*: 27–44.

Vijg, J. (1996). DNA and gene expression. In: J. E. Birren (Ed.), *Encyclopedia of Gerontology*, Vol. 1 (pp. 441–453). San Diego: Academic Press.

Vittorini, S., Paradiso, C., Donati, A., Cavallini, G., Masini, M., Gori, Z., Pollera, M., & Bergamini, E. (1999). The age-related accumulation of protein carbonyl in rat liver correlates with the age-related decline in liver proteolytic activities. *J. Gerontol.: Biol. Sci. 54A*: B318–B323.

Volicer, L., West, C. D., Chase, A. R., & Greene, L. (1983). Beta-adrenergic sensitivity in cultured smooth muscle cells: Effect of age and dietary restriction, *Mech. Ageing Dev. 21*: 283–293.

Ward. W. F. (1988). Food restriction enhances the proteolytic capacity of rat liver. *J. Gerontol.: Biol. Sci. 43*: B121–B124.

Warner, H. R. (1999). Apoptosis: A two-edged sword in aging. *Ann. N.Y Acad. Sci. 887*: 1–11.

Warner, H. R., Fernandes, G., & Wang, E. (1995). A unifying hypothesis to explain the retardation of aging and tumorigenesis by caloric restriction. *J. Gerontol.: Biol. Sci. 50A*: B107–B109.

Weindruch, R., Cheung, M. K., Verity, M. A., & Walford, R. L. (1980). Modification of mitochondrial respiration by aging and dietary restriction. *Mech. Ageing Dev. 12*: 375–392.

Weindruch, R., Kayo, T., Lee, C-K, & Prolla, T. A. (2001). Microarray profiling of gene expression in aging and its alteration by caloric restriction in mice. *J. Nutrition 131*: 918S–923S.

Weindruch, R., Kayo, T., Lee, C-K, & Prolla, T. A. (2002). Gene expression profiling using DNA microarrays. *Mech. Ageing Dev. 123*: 177–193.

Weraachakul, N., Strong, R., Wood, W. G., & Richardson, A. (1989). The effect of aging and dietary restriction on DNA repair. *Exp. Cell Res. 181*: 197–204.

Wolf, N. S. & Pendergrass, W. R. (1999). The relationships of animal age and caloric intake to cellular replication *in vivo* and *in vitro*: A review. *J. Gerontol.: Biol. Sci. 54A*: B502–B517.

Wolf, N. S., Penn, P. E., Jiang, D., Fei, R. C., & Pendergrass, W. R. (1995). Caloric restriction: conservation of *in vivo* cellular replicative capacity accompanies life span extension in mice. *Exp. Cell Res. 217*: 317–323.

Xu, X. & Sonntag, W. F. (1996a). Moderate caloric restriction prevents the age-related decline in growth hormone receptor signal transduction. *J. Gerontol.: Biol. Sci. 51A*: B167–B174.

Xu, X. & Sonntag, W. F. (1996b). Growth hormone-induced nuclear translocation of stat-3 decreases with age: modulation by caloric restriction. *Am. J. Physiol. 271*: E903–E909.

Youngman, L. D., Park, J-Y, & Ames, B. N. (1992). Protein oxidation associated with aging is reduced by dietary restriction of protein or calories. *Proc. Natl. Acad. Sci., USA 89*: 9112–9116.

Yu, B. P. (1995). Putative interventions, In: E. J. Masoro (Ed.), *Handbook of Physiology*, Section 11, *Aging* (pp. 613–631). New York: Oxford University Press.

Yu, B. P., Suescun, E. A., & Yang, S. Y. (1992). Effect of age-related lipid peroxidation on membrane fluidity and phopholipase A_2: modulation by dietary restriction. *Mech. Ageing Dev. 65*: 17–23.

Zhang, Y. & Herman, B. (2002). Ageing and apoptosis. *Mech. Ageing Dev. 123*: 245–260.

Zhen, X., Uryu, K., Cai, G., Johnson, G. P., & Friedman, E. (1999). Age-associated impairment in brain MAPK signal pathways and effect of caloric restriction in Fischer 344 rats. *J. Gerontol.: Biol. Sci. 54A*: B539–B548.

Zs-Nagy, I. (1979). The role of membrane structure and function in cellular aging: a review. *Mech. Ageing Dev. 9*: 237–246.

CHAPTER 4

Organismic physiology

Contents

Given CR's many effects on cellular and molecular processes, it would not be surprising that CR strongly influences organ and organ-system's physiology as well as integrated organismic functions. As a matter of fact, CR has been found to broadly affect organismic physiology, and its effects include, but are not limited to, the modulation of age-associated changes (Masoro, 2001).

Body composition

Not unexpectedly, body weight is reduced by CR. Sprott and Austad (1996) summarized the literature on the life span effects of a 40% reduction in food intake on body weight in a spectrum of rat and mouse strains; they found that during much of the life span, the body weight is about 40% below that of the rodents fed *ad libitum*. A graphic record of the body weight over the lifetime of *ad libitum*-fed and CR male F344 rats studied by Yu et al. (1982) is presented in Figure 4-1. Moreover, body weight by 12 months of age is similar (i.e., about 40% below that of the

RESEARCH PROFILES IN AGING
VOLUME 1 ISSN 1567-7184

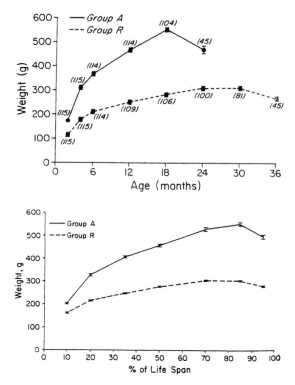

Figure 4-1. Body weight over the life span of male F344 rats fed *ad libitum* (Group A) and CR rats (Group R) restricted to 60% of the intake of Group A starting at 6 weeks of age. Upper graph plots weight against age of the rats in months (numbers in parentheses are the number of rats alive at that chronologic age); lower graph plots weight against percent of the life span of each rat. (From *J. Gerontol. 37:* 130–141, 1982; Copyright © The Gerontological Society of America. Reproduced by permission of the publisher.)

ad libitum-fed rats) whether CR (60% of *ad libitum* intake) is initiated at 6 weeks of age (soon after weaning) or at 6 months of age (Yu et al., 1985). CR also causes a reduction in body weight in rhesus monkeys (Bodkin et al., 1995; Colman et al., 1998; Kemnitz et al., 1994; Lane et al., 1995).

CR, a 40% reduction in food intake, also decreases the lean body mass of rats, but the percent reduction is somewhat less than that of body weight (Yu et al., 1982); the mass of most organs (e.g., liver, heart, kidneys, lungs, skeletal muscle) is decreased by CR, but there is no reduction in the mass of the brain or testes (Yu et al., 1982, 1984). In rhesus monkeys, a 30% reduction in food intake causes a small but significant reduction in lean body mass (Colman et al., 1998, 1999; Lane et al., 1997b).

CR markedly reduces the body fat content of rats and mice (Bertrand et al., 1980; Garthwaite et al., 1986; Harrison et al., 1984); i.e., rodents on CR are much leaner than those allowed to eat *ad libitum*. Bertrand et al. (1980) found that this reduction in fat content is due to a lifelong decrease in both the size of individual adipocytes and the number of adipocytes in the fat depots. The effects of age and CR on adipocyte number of the epididymal and perirenal fat depots of male F344 rats are shown in Figure 4-2. These findings on adipocyte number extend the work of Oscai et al. (1972), who showed that CR decreases adipose tissue cellularity in young growing rats. More recently, Barzilai and Gupta (1999) have reported that CR is particularly effective in decreasing visceral fat in rats.

Long before the effect of CR on fat mass was measured, Berg and Simms (1960) hypothesized that CR's life-prolonging action was due to a reduction in body fat content. They based this view on the fact that excess body fat is associated with premature death in humans, and it seemed likely that CR would reduce body fat content in rodents. Although the latter has since been clearly established (Bertrand et al., 1980), this hypothesis has lost favor because of a lack of correlation

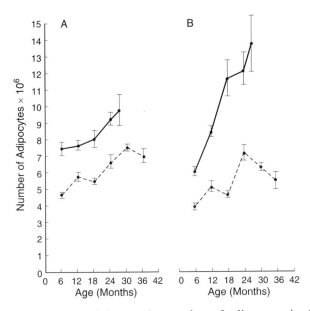

Figure 4-2. Effects of age and CR on the number of adipocytes in the epididymal depot (panel A) and the perirenal depot (panel B) of male F344 rats. The solid lines refer to the rats fed *ad libitum* and the broken lines to the CR rats. (From *J. Gerontol. 35*: 827–835, 1980; Copyright © The Gerontological Society of America. Reproduced by permission of the publisher.)

between longevity and fat mass in both rat and mouse models (Bertrand et al., 1980; Harrison et al., 1984). Nevertheless, Barzilai and Gupta (1999) feel this hypothesis should be revisited because of recent work showing that adipose tissue secretes endocrines and cytokines, and that visceral fat is a risk factor for age-associated disease in humans.

In regard to the endocrines and cytokines secreted by adipose tissue, Shimokawa and Higami (1999) reported that CR decreases the level of plasma leptin, a hardly surprising finding given the reduction in fat mass. They hypothesize that the decreased level of this modulator of the neuroendocrine system plays a key role in the anti-aging action of CR (Shimokawa & Higami, 2001). However, until a correlation is established between the reduction in the mass of specific fat depots and life extension, such hypotheses must be considered premature. Nevertheless, the "adipose tissue hypothesis" should, indeed, be revisited utilizing recently available noninvasive imaging tools.

CR also reduces the body fat content of rhesus monkeys (Colman et al., 1998, 1999; Hansen & Bodkin, 1993; Lane et al., 1997b). Moreover, CR decreases both the percentage of body fat in the abdominal region of rhesus monkeys (Figure 4-3) and the plasma concentration of leptin (Colman et al., 1999; Ramsey et al., 2000a). It also decreases the abdominal fat in cynomolgus monkeys (Cefalu et al., 1997).

The early study of McCay et al. (1935) indicated that CR has a harmful effect on bone; they reported that the femur of rats on CR is

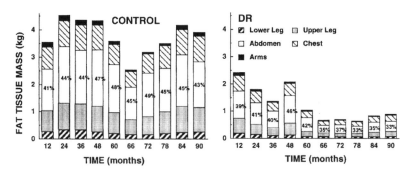

Figure 4-3. Effects of CR on regional fat mass in adult rhesus monkeys. The stacked bars represent different body regions as denoted in the right graph. The right graph designated DR refers to the monkeys on CR and the *x*-axis denotes the time after the start of CR. The left graph designated CONTROL refers to the monkeys not on CR and the *x*-axis denotes the time after the monkeys in the right graph were started on CR. (From Ramsay et al., 2000a.)

very fragile. It seems likely that the extreme level of food restriction in that study, as well as the likelihood that the rats suffered from calcium deficiency, contributed to this finding. Kalu et al. (1984b) found that when a 40% reduction in food intake is begun in 6-week-old male F344 rats, there is delayed development of a smaller but otherwise normal skeleton. The effect of CR on the femur of the male F344 rats is presented in Figure 4-4. When CR is initiated in adult rats, there is a loss of bone mass (Banu et al., 2001; Sanderson et al., 1997); however, this loss appears to be secondary to the decrease in body weight (Sanderson et al., 1997). Moreover, CR prevents senile bone loss and the marked increase in circulating parathyroid hormone, both of which occur late in life in male F344 rats (Kalu et al., 1984a); this action of CR was found to be, in part, secondary to CR's retardation of chronic renal disease (Kalu et al., 1988).

Long-term CR reduces the bone mass in rhesus monkeys, and this also appears to be secondary to reduction of body mass, specifically lean body mass (Black et al., 2001). Biochemical markers of bone turnover and hormonal regulators of bone metabolism are not affected by long-term CR in rhesus monkeys (Black et al., 2001).

Nervous system

Research on the effect of CR on the nervous system has yielded conflicting results. Some studies have shown that CR markedly retards age-associated changes, while other investigations found it has little effect.

Research on the effects of aging and CR on cognition illustrates this discord. Several studies reported that CR retards or delays age-associated decline of maze and spatial memory performance of rats and mice (Algeri et al., 1991; Goodrick, 1984; Idropo et al., 1987; Ingram et al., 1987; Pitsikas & Algeri, 1992; Stewart et al., 1989). Magnuson (1998) found that retardation of the age-associated impairment in spatial memory performance in mice correlates with the influence of CR on the densities of α-amino-3-hydroxy-5-methyl-4-isox-azolepropionate (AMPA) receptors in the frontal and parietal cortices and the CA1 and CA3 regions of the hippocampus, as well as the N-methyl-D-aspartate (NMDA) subtype of glutamate receptors in the frontal cortex and CA1 region of the hippocampus. CR's enhancement of learning and memory in old mice appears to relate to an increase in expression of the mRNA of the ε–1 subunit of the NMDA receptor and of the mRNA of the ζ–1 or ε–2 subunit (Magnuson, 2001). CR also protects rodents against

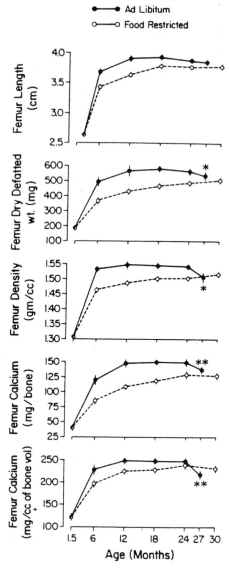

Figure 4-4. Effects of aging and CR on the femur of male F344 rats. CR is denoted by food restricted. (From Kalu et al., 1984.)

age-associated deficits in long-term potentiation (Eckles-Smith et al., 2000). Dubey et al. (1996) found that CR retards the age-associated decline in sensorimotor coordination in mice; and in aged mice, it improves performance of coping with an avoidance-learning problem.

CR initiated at 4 months of age in F344 rats improves motor learning at 14 and 22 months of age compared to *ad libitum*-fed rats of the same ages, and this improvement is associated with enhanced cerebellar noradrenergic function (Gould et al., 1995). CR prevents the age-associated loss in the ability of rats to both learn and to utilize memory (Pitsikas et al., 1990). Means et al. (1993) found that when CR is initiated as late as 14 months of age in C57BL/6 mice, the age-associated loss in cognitive functions is delayed.

However, a number of investigators have found CR ineffective in retarding cognitive decline in rats and mice. Beatty et al. (1987) found that while CR enhances the initial rate of adaptation to the radial maze, it does not improve attainment of accurate spatial memory. Bond et al. (1989) reported that CR does not prevent loss of memory at advanced ages in male Wistar rats. Campbell and Gaddy (1987) observed that CR does not retard age-associated loss in sensorimotor function of male F344 rats. Markowska (1999) also found that CR does not influence cognitive aging in male F344 rats, as indicated by two memory tests, spatial place discrimination, and repeated acquisition; at all ages, CR rats did outperform *ad libitum*-fed rats in sensorimotor tasks, though there was no change in the rate of age-associated decline in such functions. On the other hand, Markowska and Breckler (1999) did find that long-term CR improves sensorimotor functions and the place discrimination aspect of cognition in 30-month-old female F344 × BNF1 rats.

Motor function coordination declines with increasing age in mice, and CR attenuates this age change (Ingram et al., 1987). Also, CR increases locomotor activity in both rats and mice (Duffy et al., 1997).

CR has also been found to modulate age changes in nervous system morphology. In rats, neurons of the myenteric plexus of the small intestine undergo extensive age-associated cell death, starting at 12–13 months of age (Cowen et al., 2000); but CR initiated at 6 months of age protects these neurons from age-associated cell death. Also, rats show an age-associated loss in dendritic spines in layer V of parietal cortex, which is decreased by CR (Moroi-Fetters et al., 1989). CR decreases the age-associated deposition of lipofuscin in the neurons of the hippocampus and frontal cortex of C57BL/6 mice (Idropo et al., 1987). O'Steen and Landfield (1991) found that CR does not influence age-associated changes in retinal morphology in male F344 rats. In contrast, Obin et al. (2000) found that it retards the age-associated decrease in thickness of retinal layers and the decrease in density of retinal cells in BN rats. In this regard, it should be noted that the F344 rat is an albino and the BN rat is not.

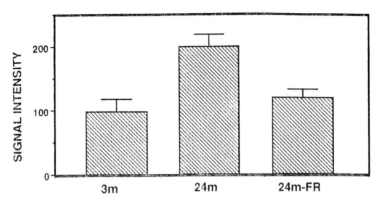

Figure 4-5. Effects of aging on astrocyte GFAP transcription in astrocytes of the hippocampal hilus of male rats. The 3m and 24 m refer to *ad libitum*-fed rats aged 3 and 24 mos. respectively and 24-FR refers to 24-mos.-old CR rats. (From Morgan et al., 1997.)

CR also influences the cellular physiology and biochemistry of neurons and glial cells. It attenuates the age-associated increase in the transcription of the glial fibrillary acidic protein gene (GFAP) in rat astrocytes (Figure 4-5), and it also decreases microglial activation during aging (Major et al., 1997; Morgan et al., 1997, 1999). CR protects the synaptic function of neurons by increasing the resistance of synaptic terminals to oxidative impairment of membrane glucose and glutamate transport, enhancing mitochondrial function as well as increasing the local levels of HSP-70 and GRP-78 (Guo et al., 2000). With age, protein carbonyl content increases and creatine kinase activity decreases in the cerebral cortex, cerebellum, and hippocampus of the BN rat; CR prevents these age changes and, in addition, it increases the activity of glutamine synthetase (Askenova et al., 1998). CR lessens the age-associated increases in microviscosity, cholesterol:phospholipid ratio, and spingomyelin:phosphatidylcholine ratio of the membranes of rat cerebral cortex (Tacconi et al., 1991). The overflow of dopamine from the striatum, which can be evoked by potassium ions in 26–28-month-old male F344 rats, is much greater for rats on long-term CR than for those fed *ad libitum* throughout life (Diao et al., 1997). CR retards the age-associated loss of dopaminergic receptors in the striatum of Wistar rats (Levin et al., 1981; Roth et al., 1984); these findings are summarized in Figure 4-6. Life-long CR enhances the rotational and stereotypic responses of old Wistar rats to intrastriatal dopamine, amphetamine, and atropine, which suggests that CR preserves the functional integrity of the striatal

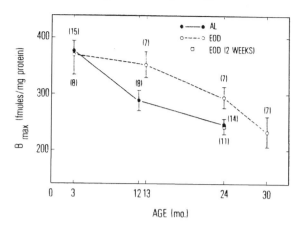

Figure 4-6. Effects of age and CR on dopamine receptor concentrations in the corpora striata of male Wistar rats. AL denotes *ad libitum*-fed rats; EOD denotes rats on long-term CR; EOD (2 weeks) denotes 24-month-old rats that have been on CR for 2 weeks. (From Roth et al., 1984.)

dopaminergic/cholinergic circuits during aging (Joseph et al., 1983). At 24 months of age, the density of striatal muscarinic binding sites is higher in male Wistar rats that have been on CR since shortly after weaning than in *ad libitum*-fed rats of the same age; and the CR rats also have higher cerebellar, hippocampal, and striatal choline acetyltransferase activity (London et al., 1985). After a brief period of CR (14 days), mRNA levels of opioid peptides are decreased in the arcuate nucleus of rodents (Kim et al., 1996).

CR also protects rodents from experimentally induced neurodegeneration. In rats, the administration of 1-methyl-4-phenyl-1,2,3,6-tetrahydropyridine (MPTP) generates a model of Parkinson's disease (an age-associated disease in humans). CR protects the dopaminergic neurons of the substantia nigra from the toxic effects of MPTP; specifically, MPTP-induced deficits in motor function as well as loss of neurons are ameliorated in 7-month-old mice on a CR regimen for 3 months, compared to *ad libitum*-fed mice of the same age (Duan & Mattson, 1999). In the same study, CR was shown to induce the expression of heat shock protein 70 and glucose-regulated protein 78; the protective actions of CR may be due to increased levels of these proteins.

The hippocampus and striatum of rats are particularly prone to damage with increasing age. Bruce-Keller et al. (1999) found that 2–4 months of CR in young rats increases the resistance of hippocampal neurons to excitotoxin-induced degeneration, and it protects striatal neurons from degeneration induced by mitochondrial toxins, 3-nitropropionic acid and malonate. In addition, CR increases the resistance of the young rats to kainic acid-induced deficits in water-maze learning and memory tasks, and, at the same time, enables them to withstand the impairment in motor function caused by 3-nitropropionic acid. CR also protects against damage resulting from transient occlusion of the middle cerebral artery in the rat stroke model (Yu & Mattson, 1999).

In mice models with a presenilin mutation, CR decreases the death of hippocampal neurons (Zhu et al., 1999). CR increases the level of brain-derived neurotrophic factor (BDNF) in hippocampal and cortical neurons (Duan et al. 2001; Lee et al., 2000), and there is evidence that this increase is linked to the protective effects of this dietary regime (Duan et al., 2001). In fact, Mattson et al. (2001) propose that CR's beneficial actions on brain are due to its ability to stimulate the expression of neurotrophic factors and stress proteins, thereby protecting neurons by suppressing reactive oxygen molecule production, stabilizing cellular calcium homeostasis, and inhibiting apoptosis. Indeed, it has been shown that CR decreases oxidative stress in many regions of mouse brain (Dubey et al., 1996). However, CR does not protect against the progression of amyotropic lateral sclerosis in a mouse model of that disease process (Pedersen & Mattson, 1999).

In summary, the evidence is compelling that CR broadly retards age-associated nervous system deterioration and also protects from neurodegeneration due to harmful agents. Nevertheless, credible studies have provided examples of CR not influencing the age-associated deterioration of a neural process.

Locomotion and skeletal muscle

Aging alters locomotor activity in humans and animal models. A longitudinal study of spontaneous locomotor activity of male F344 rats (Figure 4-7) showed that those fed *ad libitum* and those on a CR regimen exhibit a similar activity at 6 months of age; with increasing age, locomotion decreases in the *ad libitum*-fed rats but not in those on the CR regimen (Yu et al., 1985). Duffy and his colleagues have reported similar findings for male and female F344 rats (Duffy et al., 1989, 1990b) and male and female $B_6C_3F_1$ mice (Duffy et al., 1990a). When provided with a running wheel, Wistar rats on a CR regimen are less active than those fed *ad libitum* when young, but more active at older ages (Goodrick et al., 1983). Similar findings were reported for male Long–Evans rats (Holloszy & Schechtman, 1991). In contrast, when male F344 rats are given running

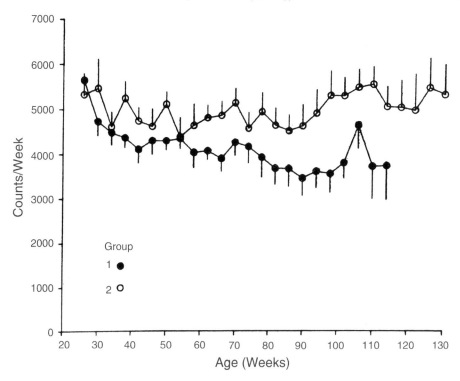

Figure 4-7. Spontaneous locomotor activity of male F344 rats fed *ad libitum* or on long-term CR. Group 1 denotes *ad libitum*-fed and Group 2 denotes CR. (From *J. Gerontol. 40*: 657–670, 1985; Copyright © The Gerontological Society of America. Reproduced by permission of the publisher.)

wheels, those on a CR regimen are much more active throughout life than *ad libitum*-fed rats (McCarter et al., 1997). Also, when mice have access to a running wheel, those on a CR regimen exhibit greater activity than those fed *ad libitum* (Ingram et al., 1987). McCarter (2000) analyzed the relationship between the effects of CR on activity level and on longevity of rats, and concluded that an increase in activity does not play a major role in the life-extension action of CR.

Studies on the effects of CR on locomotor activity of rhesus monkeys have yielded an array of conflicting findings. Kemnitz and his associates found that locomotor activity is decreased in adult rhesus monkeys after 1 year of CR (Kemnitz et al., 1993), but not significantly affected after 5 years on this dietary regimen (Ramsey et al., 1996). DeLaney et al. (1999) reported that rhesus monkeys on a CR diet for

10 years show locomotor activity similar to that of a weight-matched younger adult comparison group. Weed et al. (1997) reported increased locomotor activity associated with CR, but only in the group of rhesus monkeys in the 10–12.2 years age group and not in the 8.2–8.4 years age group. Moscrip et al. (2000) found that after 5 years on CR, 6- to 8-year-old female rhesus monkeys show a reduced locomotor activity compared to control monkeys; however, the CR regimen did not affect the locomotor activity of monkeys 10 years of age and older.

The loss of muscle mass and strength is a hallmark of the aging phenotype in both humans and animal models. Yu et al. (1982) found that from 6 to 18 months of age, the mass of the gastrocnemius muscle is less in male F344 rats on a CR regimen than in rats fed *ad libitum*. However, at ages greater than 18 months, the mass of the gastrocnemius muscle progressively decreases with age in *ad libitum*-fed rats, a decrease that does not begin in the CR rats until after 30 months of age. These findings are graphically depicted in Figure 4-8. Indeed, the effects of CR on rat skeletal muscles include: retardation of the age-associated loss of muscle mass in the soleus and extensor digitorum longus muscles of male F344 rats (Dow et al., 1988); in the soleus and anterior tibialis muscles of male Sprague-Dawley rats (El Haj et al., 1986); in the hind

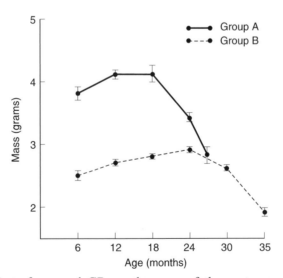

Figure 4-8. Effect of age and CR on the mass of the gastrocnemius muscles of male F344 rats. Group A denotes *ad libitum*-fed and Group B denotes CR. (From *J. Gerontol.* *37*: 130–141, 1982; Copyright © The Gerontological Society of America. Reproduced by the permission of the publisher.)

limb muscles of male BN×F344F$_1$ rats (Luhtala et al., 1994), and male Long-Evans rats (Ivy et al., 1991). In addition, CR delays the age-associated atrophy seen in the muscle fibers of the soleus, extensor digitorum longus, and flexor digitorum profundus muscles of male Sprague-Dawley rats (Boreham et al., 1988). When CR is started at 17 months of age in male Lobund-Wistar rats, it retards the age-associated loss of muscle fibers, the change in fiber type, and the increase in mitochondrial enzyme abnormalities in vastus lateralis muscles (Aspnes et al., 1997). The mass of the nonweight-bearing lateral omohyoideus muscle is less in male F344 rats on a CR regimen than in those fed *ad libitum*; however, with increasing adult age, the mass of this muscle does not decrease in either *ad libitum*-fed rats or those on a CR regimen (McCarter et al., 1982).

CR also delays age-associated skeletal muscle atrophy in mice (Bronson et al., 2000). With increasing adult age, the mitochondria of mouse skeletal muscle exhibit lipid and protein oxidative damage; CR attenuates this damage and decreases the generation of superoxide anion free radicals by the mitochondria (Lass et al., 1998).

After 18 months on a CR regimen, female rhesus monkeys have lower appendicular skeletal muscle mass than control animals of the same age; a similar finding was not seen for the male monkeys in this study (Colman et al., 1998). The influence of CR on age changes in muscle mass and function in rhesus monkeys remains to be determined.

Cardiovascular system

A loss in responsiveness to adrenergic stimulation is a major age-associated alteration in the functioning of the human heart. Kelley and Herlihy (1998) have found that the heart of the rat also exhibits a decreased response to adrenergic stimulation with increasing adult age, and long-term CR enhances the responsiveness (Table 4-1). Long-term

Table 4-1
Effects of age and CR on response of heart of male F344 Rats to isoproterenol

Age, months	nM isoproterenol conc. for 50% maximal response	
	AL	CR
4	3.7 (0.5)	1.9 (0.4)
11	5.4 (0.5)	3.2 (0.6)

Note: AL denotes *ad libitum*-fed; Langendorf isolated heart perfusion preparation; numbers in parentheses are SE. (Data are from Kelley and Herlihy, 1998.)

CR prolongs cardiac contraction and relaxation times and increases the sensitivity of the heart to isoproterenol in 10–13-month-old male F344 rats (Klebanov et al., 1997). Utilizing membranes prepared from ventricles of rats, studies show that isoproterenol-stimulated adenylate cyclase activity in the membranes decreases as the age of the donor rat increases, and that CR lessens this decline (Gao et al., 1998). In addition, CR attenuates the age-associated decrease in the ability of the cardiac synaptosomes (adrenergic nerve terminals) of male rats to release norepinephrine (Snyder et al., 1998b). It also increases the norepinephrine content of cardiac synaptosomes in adult rats by enhancing the ability to sequester norepinephrine (Kim et al., 1994; Snyder et al., 1998a).

In addition to its effects on cardiac adrenergic function, CR also affects other pump actions of rat and mouse hearts. Long-term CR prevents the reduction in myocardial contractility that occurs between 18 and 24 months of age in male F344 rats fed *ad libitum* (Chang et al., 2001). And CR of 8 months duration increases the ability of the heart of the adult male Wistar rat to maintain aortic pump flow in response to increased afterload resistance (Broderick et al., 2001). In addition, it improves recovery of heart function during reperfusion following ischemia (Broderick et al., 2001). In mice, left ventricular filling becomes increasingly dependent on atrial systole, and CR has been found to retard this age-change in cardiac function (Taffet et al., 1997). CR delays the senescent atrophy of the rat heart (Goldspink et al., 1986).

Both CR and aging are associated with alterations in cardiac molecular characteristics. CR enhances the age-associated shift in the cardiac myosin isozyme profile from the fast V_1 to the slow V_3 isoform (Klebanov & Herlihy, 1997). Baldwin and associates (Haddad et al., 1993; Morris et al., 1990) reported that severe food restriction (50–75% below the *ad libitum* intake) for 6 weeks in female Sprague-Dawley rats (age not specified) markedly influences molecular properties of the heart; unfortunately, the lack of data on the age of the rats obviates gerontologic interpretation of the findings.

Humans on a low caloric diet show decreased arterial blood pressure (Walford et al., 1992), which appears, at least in part, to be due to a suppression of sympathetic nervous system activity (Kushiro et al., 1991). CR for a period of 6 months decreases arterial blood pressure, aortic characteristic impedance, and peripheral vascular resistance in 12-month-old male Long-Evans rats, but not in 18-month-old animals (Chang et al., 1999). Long-term CR also decreases the mean arterial blood pressure of old male F344 rats and enhances the baroreflex of both young and old rats (Herlihy et al., 1992; Thomas et al., 1993). Old male

BN rats on long-term CR have a greater cerebral vascular density and blood flow than *ad libitum*-fed rats of the same age (Lynch et al., 1999).

CR inhibits the age-associated loss of force generation of aortic smooth muscle in rats (Herlihy & Yu, 1990). It also attenuates the age-associated increase in oxidative damage in aortic cells of mice (Guo et al., 2001).

Endocrines and metabolism

Insulin and carbohydrate metabolism, the growth hormone-IGF-1 axis, and the adrenal cortical glucocorticoid system will be extensively considered in Chapter 6 and, therefore, are not covered here. (Leptin and parathyroid hormone were discussed earlier in this chapter.)

Energy metabolism is a logical place to begin the discussion of endocrines and metabolism. Since animals on a CR regimen have a lower intake of energy than *ad libitum*-fed animals, their energy expenditure over any extended period of time must be decreased to a similar extent; i.e., they show a decreased metabolic rate when expressed on a per animal basis.

It is important to note that when assessing the intensity of cellular metabolic activity, metabolic rate is usually expressed per unit of body size, such as per square meter of body surface or per kg of body weight or per kg of fat-free body mass or per kg of "metabolic mass" (i.e., per $kg^{0.67}$ or per $kg^{0.75}$ of body mass). When food intake of male F344 rats is expressed per kg of body weight, those on a CR regimen have a greater daily energy intake for most of their life than those fed *ad libitum*; when expressed per kg of fat-free body mass, both dietary groups have a similar daily energy intake (Masoro et al., 1982). Not surprisingly, McCarter and Palmer (1992) reported that the metabolic rate per unit of fat-free mass or per unit of "metabolic mass" is similar over most of the lifetime for male F344 rats on CR regimen as for those fed *ad libitum* (Table 4-2). Thus these studies show that CR's anti-aging action, including the increase in longevity, in male F344 rats can occur without a decrease in metabolic rate per unit of "metabolic mass." This finding is contrary to the once widely held view that CR retards the aging processes by decreasing the intensity of energy metabolism. McCarter and McGee (1989) did find that immediately after initiating CR in male F344 rats, the metabolic rate per unit of "metabolic mass" decreases, though this decrease is a transient phenomenon with the metabolic rate returning to the level of the *ad libitum*-fed rats within a few weeks; the energy intake per unit of "metabolic mass" of these rats exhibits the same transient phenomenon (Masoro et al., 1982).

Table 4-2
Effect of age and CR on the daily metabolic rate of male F344 rats

Age, months	kcal per 24 h per kg lean body mass	
	AL	CR
6	134 (5)	141 (3)
12	124 (4)	128 (4)
18	101 (2)	113 (2)
24	138 (4)	122 (3)

Note: AL denotes *ad libitum*-fed; numbers in parentheses are SE. (Data from McCarter and Palmer, 1992.)

Duffy and his colleagues have reported similar findings on the effect of CR on metabolic rate of rats (Duffy et al., 1989) and mice (Duffy et al., 1991). The claim by Gonzales-Pacheco et al. (1993) that CR in male F344 rats decreases the metabolic rate per unit of "metabolic mass" is based on a 6-week study; Ballor (1991) reported similar findings for female Sprague-Dawley rats on a CR regime for 9 weeks. In adult Sprague-Dawley rats maintained on a CR regimen for 16 to 22 weeks, the metabolic rate is less than that of *ad libitum*-fed rats (Dulloo & Girardier, 1993); this finding is not in accord with that of McCarter's group. It is not known whether this discrepancy relates to the difference in rat strain or gender, or to the fact that CR was initiated in adult rats in the Dulloo and Girardier study and at 6 weeks of age in the studies of McCarter's group.

Greenberg and Boozer (2000) argue that expressing metabolic rate per unit of lean body mass or "metabolic mass" is inappropriate, and that it would be more accurate to express it on a per unit basis of the combined weight of the heart, kidneys, brain and liver; when they used this approach with 22-month-old male F344 rats, the CR and *ad libitum*-fed animals were found to have the same metabolic rate. However, these investigators have pointed out the need to determine the metabolic rate for each organ individually; fortunately, the technology that enables the execution of such studies is becoming available.

In two long-term studies of CR in rhesus monkeys (Lane et al., 1996; Ramsey et al., 1996, 1997), the metabolic rate per kg of lean body mass decreased following the initiation of CR, but rose to that of the nonrestricted monkeys as the CR regimen was continued. However, in another ongoing study of CR in rhesus monkeys (De Laney et al., 1999), the metabolic rate per kg lean body mass was found to be decreased after ten years on the dietary regime.

The concept that decreasing the intensity of metabolism is the mechanism by which CR increases longevity and retards aging is well entrenched, so much so that many have difficulty accepting the evidence that CR can have these effects without a decrease in the intensity of metabolism. For example, Ramsey et al. (2000b) recently attempted to resurrect the concept by suggesting that CR decreases the intensity of cellular energy expenditure; however, they have provided no credible evidence to support this view.

CR also influences protein metabolism. At most ages, isolated cells and/or cell-free systems from liver, kidney, spleen lymphocytes, and testis of rats on CR exhibit a higher rate of protein synthesis than those from animals fed *ad libitum* (Birchenall-Sparks et al., 1985; Ricketts et al., 1895). Studies with a perfused rat liver preparation showed that CR enhances both hepatic protein synthesis (Ward, 1988a) and protein degradation (Ward, 1988b) over most of the life span. These findings suggest that CR increases the rate of liver protein turnover during the greater part of the life span. Indeed, Lewis et al. (1985) reported that CR increases whole-body protein turnover in rats from 1 year of age on. However, other *in vivo* studies revealed that protein turnover is not increased by CR in all organs of Sprague-Dawley rats; e.g., it increases protein turnover in the small intestine (Merry et al., 1992) and heart ventricular muscle (Goldspink et al., 1986), but not in skeletal muscle (Goldspink et al., 1987). An increased rate of protein turnover may be a factor in slowing the rate of aging by preventing cellular accumulation of damaged proteins (Tavernakis & Driscoll, 2002).

CR markedly decreases plasma triglyceride levels in rats (Choi et al., 1988; Liepa et al., 1980; Masoro et al., 1983; Reaven & Reaven, 1981) and in rhesus monkeys (Edwards et al., 1998; Verdery et al., 1997). It also delays the age-associated increase in plasma total cholesterol, HDL-cholesterol and phospholipid levels in rats (Choi et al., 1988; Liepa et al., 1980; Masoro et al., 1983). In the rhesus monkey, CR does not influence the level of plasma total cholesterol, LDL-cholesterol, or HDL-cholesterol. However, it does decrease the molecular weight of LDL and the binding of LDL to arterial proteoglycans (Edwards et al., 1998) and lowers the plasma concentration of lipoprotein (a) in male but not in premenopausal female rhesus monkeys (Edwards et al., 2001). Also, in rhesus monkeys, it increases the level of HDL_{2b}, the sub-fraction of HDL associated with the protection from atherosclerosis in humans (Verdery et al., 1997). Based on their study of hepatic enzymes involved in intermediary metabolism in rats, Feuers et al. (1989) concluded that CR limits hepatic lipid metabolism in that species. In accord with this view,

CR was found to lower hepatic levels of triglycerides and cholesterol in rats (Choi et al., 1988) and to prevent the late-life increase in plasma ketone bodies in rats (Masoro et al., 1983). However, it must be noted in disagreement that CR increases the ability of rat liver to synthesize cholesterol from mevalonate (Choi et al., 1988).

During the Biosphere 2 project, the men and women were on a low-calorie intake for two years; these subjects exhibited a decrease in plasma total triglycerides, total cholesterol, LDL-cholesterol, and HDL-cholesterol, though the HDL_2-cholesterol increased in some participants (Verdery & Walford, 1998).

The fact that thyroid hormone markedly influences energetics, as well as many other aspects of metabolism, has led to several studies on the influence of CR on thyroid function. Armario et al. (1987) reported that a 35% reduction in caloric intake for 35 days depresses the secretion of thyroid stimulating hormone (TSH) in male Sprague-Dawley rats. More recently, a 40% reduction in caloric intake by young male F344 rats for 6 weeks was shown to result in markedly lower anterior pituitary levels of mRNA coding for TSH (Han et al., 1998). In line with these findings, a 50% reduction in caloric intake for 8 weeks markedly decreases plasma levels of thyroxine (T_4) and triiodothyronine (T_3) in male Sprague-Dawley rats (Dillmann et al., 1983). Merry and Holehan (1988) reported similar findings in a long-term CR study on this rat strain. Quigley et al. (1990) studied the effect of CR on the plasma concentrations of free T_3 and T_4 in male Sprague-Dawley rats, and they found that both were decreased after 10 weeks of a 50% reduction in caloric intake. In contrast, Swoap et al. (1994) reported that 5 weeks of a 50% reduction in caloric intake did not affect the plasma T_3 and T_4 levels in female Sprague-Dawley rats, though it markedly reduced the level of the cardiac nuclear thyroid receptor in these rats. Herlihy et al. (1990) studied the mean 24-hour levels of plasma T_3 and T_4 in male F344 rats after 19 weeks of a 40% reduction in caloric intake, and they found a significant decrease in T_3 but no effect on T_4 (Figure 4-9). Again in contrast, DeLaney et al. (1999) reported that in rhesus monkeys on a CR regime for approximately 10 years, the plasma level of T_4 is decreased while that of T_3 is not significantly altered. Ramsey et al. (2000a) also found that long-term CR in rhesus monkeys does not cause a decrease in the plasma levels of T_3. However, Mattison et al. (2001) reported that preliminary results from their long-term study of rhesus monkeys indicate that CR decreases serum T_3 levels, while not affecting serum T_4 levels.

The pancreatic islet hormone glucagon is an important regulator of hepatic metabolism; and the functional aspects of its interactions with rat

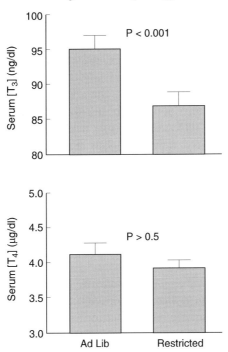

Figure 4-9. Effect of CR on the 24-h mean serum concentration of T_3 (upper panel) and T_4 (lower panel) in approximately 6-month-old male F344 rats. Ad Lib refers to *ad libitum*-fed and Restricted to CR. (From Herlihy et al., 1990.)

liver cells have been found to change with increasing age. In female WAG/Rij rats, CR attenuates the increase with age in the expression of the glucagon receptor in liver and decreases the age-associated increase in glucagon-induced hepatic accumulation of c-AMP (Teillet et al., 2002).

Both serum levels and nocturnal secretion of melatonin are known to decrease with age in both rats and humans, and it has been hypothesized that this decrease plays a causal role in senescence (MacGibbon, 1996). To test this hypothesis, MacGibbon et al. (2001) measured the excretion of a metabolite of melatonin in the urine of male Wistar rats to serve as an index of total nocturnal secretion of melatonin. They found that while CR does not prevent the age-associated decrease in melatonin secretion, the secretion of melatonin per unit of body weight is greater in the rats on CR than those fed *ad libitum*. Therefore, they believe it is likely that CR elevates the concentration of melatonin in the tissues of the rat. Two studies have found that CR increases the serum concentration of melatonin in rats (Everitt et al., 1995; Stokkan et al., 1991);

however, the meaning of these findings is questionable since the level of melatonin was measured at only one point in time during the night. Roth et al. (2001) found that peak plasma melatonin levels decrease with increasing age in rhesus monkeys, and that a 30% decrease in caloric intake over a 12-year-period prevents the age-associated decrease.

Plasma levels of dehydroepiandrosterone (DHEA) and dehydroepiandrosterone sulfate (DHEAS) decrease with increasing age in both men and women (Orentreich et al., 1992). Rhesus monkeys also exhibit an age-associated decline in the plasma concentration of DHEAS, and CR slows this decline (Lane et al., 1997a). These findings and those of several rodent studies have led to the suggestion that decreased levels of DHEA and DHEAS may play a significant role in senescence (Nelson, 1995).

Reproduction

The classic study by Osborne and Mendel (1915) demonstrated CR's profound effect on female reproduction. They started CR in female rats at 45 days of age and returned them to *ad libitum* feeding at 6 to 16 months of age; while these rats were fertile when bred between 16 and 23 months of age, none of the rats fed *ad libitum* throughout life was fertile at those ages.

In female mice, CR was found to produce a nearly sterile state, but when fed *ad libitum* from age 240 days on, they produced 13 times more litters after age 240 days than mice fed *ad libitum* throughout life were capable of (Ball et al., 1947). However, it must be pointed out that the female mice that had been on CR were unable to successfully wean their litters (Vischer et al., 1952).

Since these early studies, the effects of CR on reproduction have been explored in several careful and penetrating investigations that warrant further discussion. Berg (1960) reported that when female rats are food restricted (a 33% reduction in food intake) soon after weaning, they are able to reproduce in the age range of 730–790 days if provided food *ad libitum* though none of the rats fed *ad libitum* throughout life is able to reproduce at these ages. Female Sprague-Dawley rats on CR since weaning (about 50% reduction in calorie intake) are fertile, but their fertility during young adult life is reduced compared to the *ad libitum*-rats (Merry & Holehan, 1979); however, the rats on CR remain fertile at much older ages than fully fed rats. Puberty is delayed some 30 to 150 days in Sprague-Dawley female rats maintained on CR (about a 50% restriction) since weaning (Holehan & Merry, 1985b). Puberty is also delayed by CR in female Wistar rats (Lintern-Moore & Everitt, 1978); a 50% reduction in food intake from ages 21 to 42 days in these female Wistar rats reduces oocyte atresia, resulting in an increase in the number of oocytes per ovary at 42 days of age. Quigley et al. (1987) reported that when food intake of 4-month-old female rats is restricted by 50% for 10 weeks and then *ad libitum* feeding is resumed, there is a delay in the age-associated decline in estrous cycling, and that a similar treatment of 15- to 16-month-old rats restores estrous cycling. This ability of CR to markedly delay reproductive senescence in female rats is a robust finding that continues to be confirmed (Keenan et al., 1995).

Male Sprague-Dawley rats maintained on similar CR (about a 50% reduction in caloric intake) exhibit a delay in puberty of about 20 days (Merry & Holehan, 1981); however, CR does not affect the age-associated decrease in male fertility. When reduction in food intake of up to 30% is started at an adult age and lasts for 17 weeks, CR has little effect on the reproductive function of male and female Sprague-Dawley rats during those 17 weeks (Chapin et al., 1993).

A 40% reduction in food intake, started at 7 weeks of age in female Sprague-Dawley rats, lengthens the estrous cycle during the first 3 months but does not affect the cycle length after that; it also delays the age-associated cessation of estrous cycles (McShane & Wise, 1996). In addition, it increases the mean plasma luteinizing hormone (LH) concentration and its secretion by the adenohypophysis; this led McShane and Wise to suggest that the effects of CR on the reproductive longevity of female rats may be due to the pituitary's enhanced secretion of LH. They further suggest that CR's effect on the neuropeptide Y-pituitary axis may be involved. Indeed, there is evidence that the expression of neuropeptide Y, which augments the response of the adenohypophysis to gonadotropin-releasing hormone, is greater in the hypothalamus of rats and sheep on CR than fully fed animals (Brady et al., 1990; McShane et al., 1993). On the other hand, Holehan and Merry (1985a) reported somewhat different findings on the effect of long-term CR (about a 50% reduction in food intake) on reproductive hormones in sexually mature female Sprague-Dawley rats, namely, decreased amplitudes of peak concentrations of LH and estradiol-17β; decreased serum progesterone level; and increased level of serum FSH. However, during the period from weaning until puberty in this rat strain, CR decreases serum levels of FSH and progesterone and it increases serum levels of estradiol-17β (Holehan & Merry, 1985b). Also, CR for 3 weeks markedly decreases serum prolactin levels in female rats (Leung et al., 1983).

In summary, in both genders of rats, acutely adverse effects of CR on reproduction are modest. While in the female rat, the ability of CR to retard reproductive senescence is robust, this does not seem to be the case in male rats.

In contrast in mice, CR acutely suppresses estrous cycles (Lee et al., 1952). Indeed, as little as a 20% reduction in caloric intake between ages 3.5 and 10.5 months arrests the estrous cycle, primarily in diestrus (Nelson & Felicio, 1985). However, the follicular reserves of the CR mice are much greater than those of fully fed mice, and full feeding initiated at 10.5 months of age induces normal estrous cycling (Nelson et al., 1985). Also, CR enhances uterine function in middle-aged mice (Goodrick & Nelson, 1989). Thus, although a moderate level of CR acutely affects female reproductive function adversely in mice, it also clearly delays reproductive aging in this species. The effect of CR on the hormones of reproduction has not been appreciably studied in mice. However, 50% reduction in food intake for 1 month, started in mice of 1, 3, 7, or 18 months of age, has been found to lead to a marked reduction in prolactin secretion by the adenohypohysis (Koizumi et al., 1992a).

In female rhesus monkeys, Lane et al. (2001) have found that long-term CR does not influence the plasma concentrations of estradiol, follicle stimulating hormone (FSH), progesterone, or LH, nor did it influence the age-associated changes in the concentrations of estradiol and FSH. These investigators also reported that CR does not alter menstrual cycling or the age-associated decline in menstrual cycling in rhesus monkeys. In male rhesus monkeys, Roth et al. (1993) found that long-term CR does not influence plasma testosterone levels.

Body temperature regulation

Prior to the 1990s, there were studies that suggested that CR decreases body temperature (Cheney et al., 1983; Weindruch et al., 1979), but they lacked measurements of body temperature throughout the day. Subsequently, Duffy et al. (1991) carried out such measurements and reported that when mice are fed 60% of the *ad libitum* food intake for prolonged periods, the body temperature is markedly lower during most of the day, the mean 24-hour temperature dropping about 6°C. Indeed, a more severe food restriction (50% of the *ad libitum* intake) causes mice to experience daily torpor, with the body temperature dropping by more than 10°C (Koizumi et al., 1992b). CR also causes body temperature to decrease in rats, though much less markedly. When rats are fed 60% of the *ad libitum* intake, they exhibit a mean

24-hour reduction in body temperature of only about 2°C (Duffy et al., 1990b). Lane et al. (1996) reported that short-term CR lowers body temperature of rhesus monkeys by about 1.5°C, and it remained lower over the 6-year period of study. In contrast, DeLany et al. (1999) reported that CR does not affect the body temperature of rhesus monkeys.

It is to be expected that a reduction in body temperature would decrease the damage inherent in living processes, and thus may play a major role in the anti-aging action of CR. Indeed, there is some support for this view. Lowering the environmental temperature of several species of poikilotherms (and thus their body temperature) increases the life span (Finch, 1990). Also, housing mice at 30°C prevents most of the CR-induced decrease in body temperature, and attenuates some of the physiological effects of CR, such as slowing cellular proliferation (Koizumi et al., 1992b) and lymphopenia (Koizumi et al., 1993). Increasing the ambient temperature of rats also decreases the ability of CR to slow cellular proliferation (Jin & Koizumi, 1994). In addition, the ability of CR to extend the life of C57BL/6 mice is not seen when the mice are maintained in an ambient temperature that prevents the CR-induced decrease in body temperature (Koizumi et al., 1996); however, this finding is hard to interpret in a broad aging context because at that ambient temperature, CR is unable to prevent the development of lymphoma in this lymphoma-prone mouse strain. Therefore, it is difficult to know whether CR's lack of effect on longevity is due specifically to the inability to inhibit the occurrence of lymphoma or to a general loss in anti-aging action.

Other evidence indicates that the decrease in body temperature plays, at most, a minor role in CR's anti-aging action. As discussed in Chapter 2, CR has a similar life-extension action in both mice and rats, even though its effect on body temperature is much greater in mice than in rats. Although lowering the body temperature of fish extends life span, CR can extend their life span without lowering body temperature, and the life-extending actions of these two manipulations are additive (Walford, 1983).

Immune function

Many investigators have provided evidence that CR slows the age-associated deterioration of immune function in mice and rats, and much of the early work focused on proliferative responses of mouse lymphocytes. Gerbase-Delima et al. (1975) studied C57BL/6 mice of various ages to determine the effect of CR initiated early in life on the proliferative responses of isolated splenic lymphocytes *in vitro* to mitogens, such as phytohemoaglutinin (PHA) and poke weed mitogen (PWM). Early in life, the lymphocytes from mice on the CR regimen respond less vigorously in all tests than do lymphocytes from mice not on the CR regime. Likewise, Christdoss et al. (1984) found that CR reduces T-dependent-antigen-specific lymphocyte proliferation and antibody response in young animals; they attributed this to a defect in both the macrophages and T-cells in antigen processing and presentation, and/or cell proliferation. However, in the Gerbase-Delima et al. study, the lymphocytes from the CR mice outperformed those from the other dietary group by mid-life, and the superior performance continued into late life. Thus, in this strain of mouse, it appears that CR slows the maturation of the immune system, causing it to reach peak function at a later age and to retain high function until very advanced ages. Numerous similar findings have been reported for many different mouse and rat strains, although there are strain and species differences in the details of the changes in immune function with age (Pahlavani, 1998).

The research of Weindruch et al. (1979) offers a good example of differences between mouse strains: Unlike the C57BL/6J mouse, the immune responses of splenocytes from young as well as middle-aged and old B10C3F$_1$ mice are enhanced by CR. These investigators also reported that B-cell responses to mitogens are less affected by aging and CR than are T-cell responses, and that CR slows the rate of thymic growth and delays its involution. The natural killer (NK) cell activity is much lower in CR mice than in control mice (Weindruch et al., 1983); however, after injection with an interferon-inducing agent, which boosts NK cell responses, old CR mice show the same level of response as young mice and a much higher response than non-CR mice of a similar age. Weindruch et al. (1982a) reported that when CR is begun as late as 12 months of age, it attenuates the age-associated decline in the immune responses of splenocytes in C3B10F$_1$ mice.

CR also influences the production of immunoglobulin. Based on studies utilizing cultures of salivary gland tissue from young and old C57BL/6 mice (fed *ad libitum* or restricted to 60% of the *ad libitum* intake), it appears that CR offsets age changes in immunoglobulin production (Jolly et al., 1999).

Much effort has been focused on determining the cellular and molecular basis of the effects of CR on age changes in the immune function of mice. The ability of CR to prevent the age-associated decrease in mouse T-cell proliferation does not appear to be linked to enhanced efficiency of transmembrane signaling (Grossman et al., 1990). Weindruch et al. (1982b) reported that CR increases the proportion of PHA-responsive T-cells in the spleen of C3B10RF$_1$ mice. When begun at weaning, CR enables B6CBAT6F1 mice to retain a high level of T-cell precursors in both the cytoxic and helper lineages (Miller & Harrison, 1985). Cells producing IL-2 decrease with increasing age and CR blunts this decline (Miller, 1991). CR also prevents the increase in memory T-cells with age in B6D2F$_1$ mice, and it enables these mice to maintain both a higher number of naïve T-cells and a higher level of IL-2 production (Venkatramen et al., 1994). In B6CBAT6F1 mice, the ability of CR to maintain high levels of circulating naive T-lymphocytes at advanced ages correlates with its ability to preserve immature T-cell precursors in the thymus (Chen et al., 1998). When CR (25% reduction in food intake) is started at 12 months of age in C57BL/6 mice, it decreases the percentage of CD4$^+$ and CD8$^+$ splenocytes, reduces expression of *c-myc* by these cells, and attenuates the age-associated increase in the level of IL-6 (Volk et al., 1994).

The rat has also been studied in regard to the molecular and cellular basis of the effects of CR on age changes in immune function. CR regimens retard the age-associated decline in calcium ion content and membrane fluidity and blunt the increase in the arachadonic acid content of the phospholipids in splenocytes of F344 rats (Fernandes et al., 1990). As seen with the mouse, CR increases interleukin-2 (IL-2) production by lymphocytes of F344 rats (Pahlavani et al., 1990; Venkatraman & Fernandes, 1992), and the cells from the CR rats show increased levels of IL-2 mRNA (Pahlavani et al., 1990). When nuclear extracts of T-cells from CR rats are treated with concanavalin A (Con A), they exhibit an increase in the transcription factor NFAT binding activity, compared to *ad libitum*-fed rats; this increase correlates with an increase in IL-2 gene expression (Pahlavani et al., 1997). However, Goonewardene and Murasko (1995) did not observe an effect of CR on IL-2 expression in lymphocytes of 30-month-old BN rats, again indicating strain differences in the response of the immune system. CR enhances the expression of IL-2 receptors in T cells of F344 rats (Iwai & Fernandes, 1989; Venkatraman & Fernandes, 1992). In the Lobund-Wistar rat, CR influences the makeup of the lymphocyte population by increasing the percentage of following cellular subsets: T-cells (CD3$^+$), cytoxic/suppressor T-cells (CD8$^+$), and natural killer (NK) cells (OX8$^+$ OX19$^-$) (Gilman-Sachs et al., 1991). However, CR does not influence NK cell responses in middle-aged and old rats (Riley-Roberts et al., 1989). It does retard the age-associated shift in lymphocytes from naïve to memory cells in F344×BNF$_1$ rats (Fernandes et al., 1997); this action may underlie the higher level of the antiinflammatory IL-2 and the lower levels of proinflammatory interleukin-6 (IL-6) and tumor necrosis factor-α (TNFα) in old rats on CR, compared to the levels in rats of the same age fed *ad libitum*.

CR attenuates the age-associated decline in the immune response of mice to the influenza-A virus (Effros et al., 1991); specifically, the decline in antigen presentation, T-cell proliferation and antibody production are attenuated. The fact that CR enhances the apoptotic elimination of non-functional T-cells in old mice may, in part, underlie its protection against influenza (Spaulding et al., 1997); the response of the splenic T cells to *in vitro* apoptotic stimuli is shown in Figure 4-10. CR also slows the progression of the immunodeficiency syndrome induced in mice by murine retrovirus (Fernandes et al., 1992).

Macrophages from 6-month-old C57BL/6 mice on a CR regime are less responsive to *in vitro* lipopolysaccharide stimulation (Sun et al., 2001), as evidenced by lower expression of IL-6 and interleukin 12 (IL-12). These mice were also tested for their response to polymicrobial sepsis induced by cecal ligation and puncture; the mice on CR had higher serum levels of proinflammatory cytokines TNFα and IL-6 and died earlier than the *ad libitum*-fed mice. Thus, in this case CR adversely affected the animal. However, these were young mice that had been on CR for only 5 months. Gerbase-Delima et al. (1975) have shown that CR slows the maturation of the immune system in this mouse strain. Therefore, it would be of great interest to know if similar findings would be obtained in old mice on long-term CR. Dong et al. (1998) reported that when alveolar macrophages from CR rats (25% reduction in food intake) are stimulated with lipopolysaccharide *in vitro,* they generate less TNFα and NO as well as lower levels of TNFα- and IL-6-mRNA than cells from *ad libitum*-fed rats.

After rhesus monkeys at the University of Wisconsin Primate Research Center were maintained on an adult-onset CR regimen for 2–4 years, they were assessed for immunologic functions; mitogen-induced proliferation of peripheral blood mononuclear cells, NK cell lysis, and plasma antibody response to influenza vaccine were all significantly reduced (Roecker et al., 1996). Rhesus monkeys maintained on CR at the NIH Primate Unit were also assessed for immunologic functions (Weindruch et al., 1997); the CR regimen was initiated at either about 1 year of age (young group) or 3–5 years of age (young adult group).

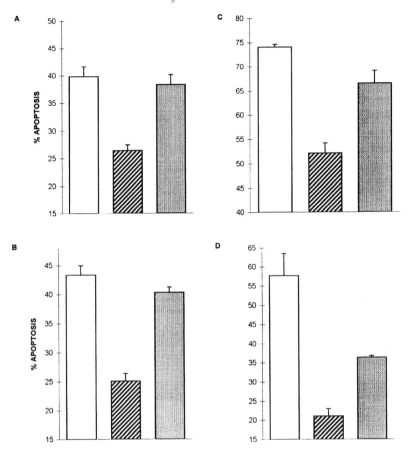

Figure 4-10. Effects of age and CR on the *in vitro* response of splenic T-cells of female C3B10F mice to apoptotic stimuli. Panel A, anti-mouse CD3; panel B, staurosporine; panel C, hyperthermia; panel D, gamma irradiation. Open bars, *ad libitum*-fed age 6–9 months; hatched bars, *ad libitum*-fed age 28–36 months; shaded bars, long-term CR age 28–36 months. (From Spaulding et al., 1997.)

Both groups had been on the CR regimen for 7 years at the time of the immunologic assessment. Compared to the normally fed controls, the mitogen-induced proliferative responses of peripheral blood mononuclear cells were reduced in the young group but not in the young adult group; CR caused lymphopenia in both groups. Clearly, research on the effects of CR on immune functions in rhesus monkeys is only beginning; therefore, firm conclusions about its actions in this species must await further study.

Wound healing

Increasing age impairs wound healing in mice, rats and rhesus monkeys and long-term CR does not retard this impairment (Harrison & Archer, 1987; Reiser et al., 1995; Roth et al., 1997). In fact, Harrison and Archer, and Reiser et al. found that long-term CR has an adverse effect on wound healing in mice and rats. In rats, the decrease is associated with decreased collagen production and decreased levels of the difunctional crosslink dihydroxylisinonorleucine (Reiser et al., 1995). On the other hand, Roth et al. reported no statistically significant effect of CR on wound healing in rats and rhesus monkeys, though they did detect a tendency for CR to enhance healing. Interestingly, a mouse study by Reed et al. (1996) indicates that CR does maintain the youthful capacity to heal wounds into old age, but this effect is manifested only with an abundant intake of energy just prior to and following the wounding. Specifically, when *ad libitum* feeding is started 4 weeks prior to wounding, old CR mice heal as rapidly as young mice fed *ad libitum* and much more rapidly than old mice fed *ad libitum* for their entire life. This increased capacity for wound repair appears to relate to an enhancement of collagen biosynthesis and cell proliferation when the CR animals are provided an abundant dietary source of energy.

References

Algeri, S., Biagini, L., Manfredi, A., & Pitsikas, N. (1991). Age-related ability of rats kept on life-long hypocaloric diet in spatial memory test; longitudinal observations. *Neurobiol. Aging 12*: 277–282.

Armario, A., Montero, J. L., & John, T. (1987). Chronic food restriction and circadian rhythms of pituitary-adrenal hormone, growth hormone, and thyroid stimulating hormone. *Ann. Nutr. Metab. 31*: 81–87.

Askenova, M. V., Askenov, M. Y., Carney, J. M., & Butterfield, D. A. (1998). Protein oxidation and enzyme activity decline in old Brown Norway rats are reduced by dietary restriction. *Mech. Ageing Dev. 100*: 157–168.

Aspnes, L. E., Lee, C. M., Weindruch, R., Chung, S. S., Roecker, E. B., & Aiken, J. M. (1997). Caloric restriction reduces fiber loss and mitochondrial abnormalities in aged rat muscle. *FASEB J. 11*: 573–581.

Ball, Z. B., Barnes, R. H., & Visscher, M. B. (1947). The effects of dietary caloric restriction on maturity and senescence with particular reference to fertility and longevity. *Am. J. Physiol. 150*: 511–519.

Ballor, D. L. (1991). Effect of dietary restriction and/or exercise on 23-h metabolic rate and body composition in female rats. *J. Appl. Physiol. 71*: 801–806.

Banu, J., Orhii, P. B., Okafor, M. C., Wang, L., & Kalu, D. N. (2001). Analysis of the effects of growth hormone, exercise, and food restriction on cancellous bone in different bone sites in middle-aged female mice. *Mech. Ageing Dev. 122*: 849–864.

Barzilai, N. & Gupta, G. (1999). Revisiting the role of fat mass in the life extension induced by caloric restriction. *J. Gerontol.: Biol. Sci. 54A*: B89–B96.

Beatty, W. W., Clouse, B. A., & Bierley, R. A. (1987). Effects of long-term restricted feeding on the radial maze performance by aged rats. *Neurobiol. Aging 8*: 325–327.

Berg, B. N. (1960). Nutrition and longevity in the rat. I. Food intake in relation to size, health and fertility. *J. Nutrition 71*: 242–254.

Berg, B. N., & Simms, H. S. (1960). Nutrition and longevity in the rat. II. Longevity and the onset of disease with different levels of intake. *J. Nutrition 71*: 255–263.

Bertrand, H. A., Lynd, F. T., Masoro, E. J., & Yu, B. P. (1980). Changes in adipose tissue mass and cellularity through adult life of rats fed ad libitum or a life-prolonging restricted diet. *J. Gerontol. 35*: 827–835.

Birchenall-Sparks, M. C., Roberts, M. S., Staecker, J., Hardwick, J. P., & Richardson, A. (1985). Effect of dietary restriction on liver protein synthesis in rats. *J. Nutrition 115*: 944–950.

Black, A., Allison, D. B., Shapses, S. A., Tilmont, E. M., Handy, A. M., Ingram, D. K., Roth, G. S., & Lane, M. A. (2001). Caloric restriction and skeletal mass in rhesus monkeys (*Macaca mulatta*): Evidence for an effect mediated through changes in body size. *J. Gerontol.: Biol. Sci. 56A*: B98–B107.

Bodkin, N. L., Ortmeyer, H. K., & Hansen, B. C. (1995). Long-term dietary restriction in older-aged rhesus monkeys: effects on insulin resistance. *J. Gerontol.: Biol. Sci. 50A*: B142–B147.

Bond, N. W., Everitt, A. V., & Walton, J. (1989). Effect of dietary restriction on radial-arm maze performance and flavor memory in aged rats. *Neurobiol. Aging 10*: 27–30.

Boreham, C. A. G., Watt, P. W., Williams, P. E., Merry, B. J., Goldspink, G., & Goldspink, D. F. (1988). Effects of aging and chronic dietary restriction on morphology of fast and slow muscles of the rat. *J. Anat. 157*: 111–125.

Brady, L. S., Smith, M. A., Gold, P. W., & Herkenham, M. (1990). Altered expression of hypothalamic neuropeptide mRNAs in food restricted and food deprived rats. *Neuroendocrin. 52*: 441–447.

Broderick, T. L., Driedzic, W. R., Gillis, M., Jacob, J., & Belke, T. (2001). Effects of chronic food restriction and exercise training on recovery of cardiac function following ischemia. *J. Gerontol.: Biol. Sci. 56A*: B33–B37.

Bronson, K. N., Lipman, R. T., Ding, W., Lamont, L., Cosmos, A. C., & Manfredi, T.G. (2000). Diet restriction and age alters skeletal muscle capillarity in B6C3F1 mice. *Age 23*: 141–145.

Bruce-Keller, A. J., Umberger, G., McFall, R., & Mattson, M. P. (1999). Food restriction reduces brain damage and improves behavioral outcome following exitotoxic and metabolic insults. *Ann. Neurol. 45*: 8–15.

Campbell, B. A. & Gaddy, J. R. (1987). Rate of aging and dietary restriction: Sensory and motor function in the Fischer 344 rat. *J. Gerontol. 42*: 154–159.

Cefalu, W. T., Wagner, J. D., Wang, Z. Q., Bell-Farrow, A. D., Collins, J., Haskell, D., Bechtold, R., & Morgan, T. (1997). A study of caloric restriction and cardiovascular aging in cynomolgus monkeys *(Macaca fascicularis)*: A potential model for aging research. *J. Gerontol.: Biol. Sci. 52A*: B10–B19.

Chang, K-C., Chow, C-Y., Peng, Y-I., Chen, T-J., & Tsai, Y-F. (1999). Effects of food restriction on mechanical properties of arterial system in adult and middle-aged rats. *J. Gerontol.: Biol. Sci. 54A*: B441–B447.

Chang, K-C., Peng, Y-I., Lee, F-C., & Tseng, Y-Z. (2001). Effects of food restriction on systolic mechanical behavior of the ventricular pump in middle-aged and senescent rats. *J. Gerontol.: Biol. Sci. 56A*: B108–B114.

Chapin, R. E., Gulati, D. K., Barnes, L. H., & Teague, J. L. (1993). The effect of feed restriction on reproductive function in Sprague-Dawley rats. *Fund. Appl. Toxicol. 20*: 23–29.

Chen, J., Astle, C. M., & Harrison, D. E. (1998). Delayed immune aging in diet-restricted B6CBAT6F1 mice is associated with preservation of naive T cells. *J. Gerontol.: Biol. Sci. 53A*: B330–B337.

Cheney, K. E., Liu, K. K., Smith, G. S., Meredith, P. J., Mickey, M. R., & Walford, R. L. (1983). The effect of dietary restriction of varying duration on survival, tumor patterns, immune function, and body temperature in B10C3F$_1$ female mice. *J. Gerontol. 38*: 420–430.

Choi, Y-S., Goto, S., Ikeda, I., & Sugano, M. (1988). Age-related changes in lipid metabolism in rats: the consequence of moderate food restriction. *Biochim. Biophys. Acta 963*: 237–242.

Christdoss, P., Talal, N., Lundstrom, J., & Fernandes, G. (1984). Suppression of cellular and humoral immunity to T-dependent antigens by calorie restriction. *Cell. Immunol. 88*: 1–8.

Colman, R. J., Ramsey, J. J., Roecker, E. B., Havighurst, T., Hudson, J. C., & Kemnitz, J. W. (1999). Body fat distribution with long-term dietary restriction in adult male rhesus macaques. *J. Gerontol.: Biol. Sci. 54A*: B283–B290.

Colman, R. J., Roecker, E. B., Ramsey, J. J., & Kemnitz, J. W. (1998). The effect of dietary restriction on body composition in adult male and female rhesus macaques. *Aging Clin. Exp. Res. 10*: 83–92.

Cowen, T., Johnson, R. J. R., Soubeyre, V., & Santer, R. M. (2000). Restricted diet rescues rat enteric motor neurons from age related cell death. *Gut 47*: 653–660.

DeLaney, J. P., Hansen, B. C., Bodkin, N. L., Hannah, J., & Bray, G. A. (1999). Long-term caloric restriction reduces energy expenditure in aging monkeys. *J. Gerontol.: Biol. Sci. 54A*: B5–B11.

Diao, L. H., Bickford, P. C., Stevens, J. O., Cline, E. J., & Gerhardt, G. A. (1997). Caloric restriction enhances evoked DA overflow in striatum and nucleus accumbens of aged Fischer 344 rats. *Brain Res. 763*: 276–280.

Dillman, W. H., Berry, S., & Alexander, N. (1983). A physiological dose of triiodothyronine normalizes cardiac myosin adenosine triphosphatase activity and changes myosin isoenzyme distribution in semistarved rats *Endocrinology 112*: 2081–2087.

Dong, W. Selgrade, M. K., Gilmour, I. M., Lange, R. W., Park, P., Luster, M. I., & Kari, F. W. (1998). Altered alveolar macrophage function in calorie-restricted rats. *Am. J. Respir. Cell Mol. Biol. 19*: 920–935.

Dow, C. K., Starnes, J. W., & White, T. T. (1988). Muscle atrophy and hypoplasia with aging: impact of training and food restriction. *J. Appl. Physiol. 64*: 2428–2432.

Duan, W. & Mattson, M. P. (1999). Dietary restriction and 2-deoxyglucose administration improve behavioral outcome and reduce degeneration of dopaminergic neurons in models of Parkinson's disease. *J. Neurosci. Res. 57*: 195–206.

Duan, W., Guo, Z., & Mattson, M. P. (2001). Brain-derived neurotrophic factor mediates an excitoprotective effect of dietary restriction in mice. *J. Neurochem. 76*: 619–626.

Dubey, A., Forster, M. J., Lal, H., & Sohal, R. S. (1996). Effect of age and caloric intake on protein oxidation in different brain regions and on behavioral functions of the mouse. *Arch. Biochem. Biophys. 333*: 189–197.

Duffy, P. H., Feuers, R., & Hart, R. W. (1990a). Effect of chronic caloric restriction on the circadian regulation of physiological and behavioral variables in old male $B_6C_3F_1$ mice. *Chronobio. Int. 7*: 291–303.

Duffy, P. H., Feuers, R., Leakey, J., & Hart, R. W. (1991). Chronic caloric restriction in old female mice: changes in circadian rhythms of physiological and behavioral variables. In: L. Fishbein (Ed), *Biological Effects of Dietary Restriction*, (pp. 245–263). Berlin: Springer-Verlag.

Duffy, P. H., Feuers, R., Leakey, J., Nakamura, K. D., Turturro, A., & Hart, R. W. (1989). Effect of chronic restriction on physiological variables related to energy metabolism in the male Fischer 344 rat. *Mech. Ageing Dev. 48*: 117–133.

Duffy, P. H., Feuers, R., Nakamura, K. D., Leakey, J., & Hart, R. W. (1990b). Effect of chronic caloric restriction on synchronization of various physiological measures in old female Fischer 344 rats. *Chronobio. Int. 7*: 113–124.

Duffy, P. H., Leakey, J., Pipkin, J. L., Turturro, A., & Hart, R. W. (1997). The physiologic, neurologic, and behavioral effects of caloric restriction related to aging, disease, and environmental factors. *Environ. Res. 73*: 242–248.

Dulloo, A. G. & Girardier, L. (1993). 24 hour energy expenditure several months after weight loss in the underfed rat: evidence for a chronic increase in whole-body metabolic efficiency. *Inter, J. Obesity 17*: 115–123.

Eckles-Smith, K., Clayton, D., Bickford, P., & Browning, M. D. (2000). Caloric restriction prevents age-related deficits in LTP and NMDA receptor expression. *Mol. Brain Res. 78*: 154–162.

Edwards, I. J., Rudel, L. L., Terry, J. G., Kemnitz, J. W., Weindruch, R., & Cefalu, W. T. (1998). Caloric restriction in rhesus monkeys reduces low density lipoprotein interaction with arterial proteoglycans. *J. Gerontol.: Biol. Sci. 53A*: B443–B448.

Edwards, I. J., Rudel, L. L., Terry, J. G., Kemnitz, J. W., Weindruch, R., Zaccaro, D. J., & Cefalu, W. T. (2001). Caloric restriction lowers plasma lipoprotein (a) in male but not female rhesus monkeys. *Exp. Gerontol. 36*: 1413–1418.

Effros, R. B., Walford, R. L., Weindruch, R., & Mitcheltree, C. (1991). Influence of dietary restriction on immunity to influenza in aged mice. *J. Gerontol.: Biol. Sci. 46*: B142–B147.

El Haj, A. J., Lewis, S. E. M., Goldspink, D. F., Merry, B. J., & Holehan, A. M. (1986). The effect of chronic and acute dietary restriction on the growth and protein turnover of fast and slow types of rat skeletal muscle. *Comp. Biochem. Physiol. 85A*: 281–287.

Everitt, A. V., Destafanis, P., Parkes, A. A., Cairncross, K. D., & Eyland, A. (1995). The effect of neonatal pinealectomy on the inhibitory action of food restriction on vaginal opening and collagen aging in the rat. *Mech. Ageing Dev. 78*: 39–45.

Fernandes, G., Flescher, E., & Venkatramen, J. T. (1990). Modulation of cellular immunity, fatty acid concentration, fluidity, and Ca^{2+} influx by food restriction in aging rats. *Aging, Immunol. Infect. Dis. 2*: 117–125.

Fernandes, G., Tormar, V., & Venkatraman, J. T. (1992). Potential diet therapy on murine AIDS. *J. Nutrition 122*: 716–722.

Fernandes, G., Venkatraman, J. T., Turturro, A., Attwood, V. G., & Hart, R. W. (1997). Effect of food restriction on life span and immune functions in long-lived Fischer 344xBrown Norway F_1 rats. *J. Clin. Immunol. 17*: 85–95.

Feuers, R. J., Duffy, P. H., Leakey, J. A., Turturro, A., Mittelstaedt, R. A., & Hart, R. A. (1989). Effect of chronic caloric restriction on hepatic enzymes of intermediary metabolism in the male Fischer rat. *Mech. Ageing Dev. 48*: 179–189.

Finch, C. E. (1990). *Longevity, Senescence, and the Genome.* Chicago: University of Chicago Press.

Gao, E., Snyder, D. L., Roberts, J., Friedman, E., Cai, G., Pelleg, A., & Horwitz, J. (1998). Age-related decline in beta adrenergic and adenosine A_1 receptor function in the heart are attenuated by dietary restriction. *J. Pharmacol. & Exp. Therap. 285*: 186–192.

Garthwaite, S. M., Cheng, H., Bryan, J. E., Craig, B. W., & Hollozy, J. O. (1986). Ageing, exercise and food restriction: Effects on body composition. *Mech. Ageing Dev. 36*: 187–196.

Gerbase-Delima, M., Liu, R. K., Cheney, K. E., Mickey, R., & Walford, R. L. (1975). Immune function and survival in the long-lived mouse strain subjected to under-nutrition. *Gerontologia 21*: 184–193.

Gilman-Sachs, A., Kim, Y. B., Pollard, M., & Snyder, D. L. (1991). Influence of aging, environmental antigens, and dietary restriction on expression of lymphocyte subsets in germ-free and conventional Lobund-Wistar rats. *J. Gerontol.: Biol. Sci. 46*: B101–B109.

Goldspink, D. F., El Haj, A. J., Lewis, S. E., Merry, B. J., & Holehan, A. M. (1987). The influence of chronic dietary intervention on protein turnover and growth of the diaphragm and extensor digitorum longus muscles of the rat. *Exp. Gerontol. 22*: 67–78.

Goldspink, D. F., Lewis, S. E. M., & Merry, B. J. (1986). Effects of aging and long term dietary intervention on protein turnover and growth of ventricular muscle in the rat heart. *Cardiovasc. Res. 20*: 672–678.

Gonzales-Pacheco, D. M., Buss, W. C., Koehler, K. M., Woodside, W. F., & Alpert, S. S. (1993). Energy restriction reduces metabolic rate in adult male Fischer-344 rats. *J. Nutrition 123*: 90–97.

Goodrick, C. L. (1984). Effects of lifelong restricted feeding on complex maze performance in rats. *Age 7*: 1–2.

Goodrick, C. L., Ingram, D. K., Reynolds, M. A., Freeman, J. R., & Cider, N. L. (1983). Effect of intermittent feeding on growth, activity and lifespan in rats allowed voluntary exercise. *Exp. Aging Res. 9*: 203–209.

Goodrick, G. J., & Nelson, J. F. (1989). The decidual cell response in aging C57BL/6J mice is potentiated by long-term ovariectomy and chronic food restriction. *J. Gerontol.: Biol. Sci. 44*: B67–B71.

Goonewardene, L. M. & Murasko, D. M. (1995). Age-associated changes in mitogen-induced lymphoproliferation and lymphokine production in the long-lived Brown-Norway rat: effect of caloric restriction. *Mech. Ageing Dev. 83*: 103–116.

Gould, T. J., Bowenkamp, K. E., Larson, G., Zahniser, N. R., & Bickford, P. C. (1995). Effects of dietary restriction on motor learning and cerebellar noradrenergic dysfunction in aged F344 rats. *Brain Res. 684*: 150–158.

Greenberg, J. A. & Boozer, C. N. (2000). Metabolic mass, metabolic rate, caloric restriction, and aging in male Fischer 344 rats. *Mech. Ageing Dev. 113*: 37–48.

Grossmann, A., Maggio-Price, L., Jinnemaun, J. C., Wolf, N. S., & Rabinovitch, P, S. (1990). The effect of long-term caloric restriction on function of T-cell subsets in old mice. *Cell. Immunol. 131*: 191–204.

Guo, Z., Ersoz, A., Butterfield, D. A., & Mattson, M. P. (2000). Beneficial effects of dietary restriction on cerebral cortical synaptic terminals: preservation of glucose transport and mitochondrial function after exposure to β-amyloid peptide and oxidative and metabolic insults. *J. Neurochem. 75*: 314–320.

Guo, Z., Young, H., Hamilton, M. L., Van Remmen, H., & Richardson, A. (2001). Effects of age and food restriction on oxidative damage and antioxidant enzymes in the mouse aorta. *Mech. Ageing Dev. 122*: 1771–1786.

Haddad, F., Bodell, P. W., McCue, S. A., Herrick, R. E., & Baldwin, K. M. (1993). Food restriction-induced transformations in cardiac functional and biochemical properties in rats. *J. Appl. Physiol. 74*: 606–612.

Han, E-S., Lu, D. H., & Nelson, J. F. (1998). Food restriction differentially affects mRNAs encoding the major anterior pituitary tropic hormones. *J. Gerontol.: Biol. Sci. 53A*: B322–B329.

Hansen, B. C. & Bodkin, N. L. (1993). Primary prevention of diabetes mellitus by prevention of obesity in monkeys. *Diabetes 42*: 1809–1814.

Harrison, D. E. & Archer, J. R. (1987). Genetic differences in effects of food restriction on aging in mice. *J. Nutrition 117*: 376–382.

Harrison, D. E., Archer, J. R., & Astole, C. M. (1984). Effects of food restriction on aging: separation of food intake and adiposity. *Proc. Natl. Acad. Sci., USA 81*: 1835–1838.

Herlihy, J. T. & Yu, B. P. (1990). Dietary manipulation of age-related decline in vascular smooth muscle function. *Am. J. Physiol. 238*: H652–H655.

Herlihy, J. T., Stacy, C., & Bertrand, H. A. (1990). Long-term food restriction depresses serum thyroid concentration in the rat. *Mech, Ageing Dev. 53*: 9–16.

Herlihy, J. T., Stacy, C., & Bertrand, H. A. (1992). Long-term caloric restriction enhances baroreflex responsiveness in Fischer 344 rats. *Am. J. Physiol. 263*: H1021–H1025.

Hishinuma, K., Nishimura, T., Konno, A., Hashimoto, Y., & Kimura, S. (1988). The effect of dietary restriction on mouse T cell functions. *Immuno. Lett. 17*: 351–356.

Holehan, A. M. & Merry, B. J. (1985a). Modification of the estrous cycle hormonal profile by dietary restriction. *Mech. Ageing Dev. 32*: 63–76.

Holehan, A. M. & Merry, B. J. (1985b). The control of puberty in the dietary restricted rat. *Mech. Ageing Dev. 32*: 179–191.

Holloszy, J. O. & Schechtman, K. B. (1991). Interaction between exercise and food restriction: Effects on longevity of male rats. *J. Appl. Physiol. 70*: 1529–1535.

Idropo, F., Nandy, K., Mostofsky, D. I., Blatt, L., & Nandy, L. (1987). Dietary restriction: Effects on radial maze learning and lipofuscin in neurons of hippocampus and frontal cortex. *Arch. Gerontol. Geriatr. 6*: 355–362.

Ingram, D. K., Weindruch, R., Spangler, E. L., Freeman, J. R., & Walford, R. L. (1987). Dietary restriction benefits learning and motor performance of aged mice. *J. Gerontol. 42*: 78–81.

Ivy, J. L., Young, J. C., Craig, B. W., Kohrt, W. M., & Holloszy, J. O. (1991). Ageing, exercise, and food restriction: effects on skeletal muscle glucose uptake. *Mech. Ageing Dev. 61*: 199–213.

Iwai, H. & Fernandes, G. (1989). Immunological functions in food restricted rats: Enhanced expression of high-affinity IL-2 receptors on splenic T cells. *Immunol. Lett. 23*: 125–132.

Jin, Y-H. & Koizumi, A. (1994). Decreased cellular proliferation by energy restriction is recovered by increasing housing temperature. *Mech. Ageing Dev. 75*: 59–67.

Jolly, C. A., Fernandez, R., Muthukumar, A. R., & Fernandes, G. (1999). Calorie restriction modulates Th-1 and Th-2 cytokine-induced immunoglobulin secretion in young and old C57BL/6 cultured submandibular glands. *Aging Clin. Exp. Res. 11*: 383–389.

Joseph, J. A., Whitaker, J., Roth, G. S., & Ingram, D. K. (1983). Life-long dietary restriction affects striatally-mediated behavioral responses in aged rats. *Neurobiol. Aging 4*: 191–196.

Kalu, D. N., Hardin, R. H., Cockerham, R., & Yu, B. P. (1984a). Aging and dietary modulation of rat skeleton and parathyroid hormone. *Endocrinology 115*: 1239–1247.

Kalu, D. N., Hardin, R. R., Cockerham, R., Yu, B. P., Norling, B. K., & Egan, J. W. (1984b). Lifelong food restriction prevents senile osteopenia and hyperparathyroidism in F344 rats. *Mech. Ageing Dev. 26*: 103–112.

Kalu, D. N., Masoro, E. J., Yu, B. P., Hardin, R. R., & Hollis, B. W. (1988). Modulation of age-related hyperparathyroidism and senile bone loss in Fischer rats by soy protein and food restriction. *Endocrinology 122*: 1847–1854.

Keenan, K. P., Soper, K. A., Hertzog, P. R., Gumprecht, L. A., Smith, P. F., Mattson, B. A., Ballam, G. C., & Clark, R. L. (1995). Diet, overfeeding, and moderate dietary restriction in control Sprague-Dawley rats: II. Effects on age-related proliferative and degenerative lesions. *Toxicol Path. 23*: 287–302.

Kelley, G. R. & Herlihy, J. T. (1998). Food restriction alters the age-related decline in cardiac β–adrenergic responsiveness. *Mech. Ageing Dev. 103*: 1–12.

Kemnitz, J. W., Roecker, E. B., Weindruch, R., Olson, D. F., Baum, S. T., & Bergman, R. N. (1994). Dietary restriction increases insulin sensitivity and lowers blood glucose in rhesus monkeys. *Am. J. Physiol. 266*: E540–E547.

Kemnitz, J. W., Weindruch, R., Roecker, E. B., Crawford, K., Kaufman, P. L., & Ehrshler, W. B. (1993). Dietary restriction of adult male rhesus monkeys: Design, methodology, and preliminary findings from the first year of study. *J. Gerontol.: Biol. Sci. 48*: B17–B26.

Kim, E-M., Welch, C. C., Grace, M. K., Billington, C. J., & Levine, A. S. (1996). Chronic food restriction and acute food deprivation decreases mRNA levels of opioid peptides in arcuate nucleus. *Am. J. Physiol. 270*: R1019–R1024.

Kim, S. W., Yu, B. P., Sanderford, M., & Herlihy, J. T., (1994). Dietary restriction modulates the norepinephrine content and uptake of the heart and cardiac synaptosomes. *Proc. Soc. Exp. Biol. Med. 207*: 43–47.

Klebanov, S. & Herlihy, J. T. (1997). Effect of life-long food restriction on cardiac myosin composition. *J. Gerontol.: Biol. Sci. 52A*: B184–B189.

Klebanov, S., Herlihy, J. T., & Freeman, G. L. (1997). Effect of long-term food restriction on cardiac mechanics. *Am. J. Physiol. 273*: H2333–H2342.

Koizumi, A. Roy, N. S., Tsukada, M., & Wada, Y. (1993). Increase in housing temperature can alleviate decreases in white cell counts after energy restriction in C57BL/6 female mice. *Mech. Ageing Dev. 71*: 97–102.

Koizumi, A., Tsukada, M., Masuda, H., Kamiyama, S., & Walford, R. L. (1992a). Specific inhibition of pituitary prolactin production by energy restriction in C3H/SHN female mice. *Mech. Ageing Dev. 64*: 21–35.

Koizumi, A., Tsukada, M., Wada, Y., Masuda, H., & Weindruch, R. (1992b). Mitotic activity in mice is suppressed by energy restriction-induced torpor. *J. Nutrition 122*: 1446–1453.

Koizumi, A., Wada, Y., Tsukada, M., Kayo, T., Naruse, M., Horiuchi, K., Mogi, T., Yoshioka, M., Sasaki, M., Miyamaura, Y., Abe, T., Ohtomo, K., & Walford, R. L. (1996). A tumor preventive effect of dietary restriction is antagonized by high housing temperature through deprivation of torpor. *Mech. Ageing Dev. 92*: 67–82.

Kushiro, T., Kobayashi, F., Osada, H., Tomizama, H., Satoh, K., Otsuka, Y., Kurumatani, H., & Kajiwara, N. (1991). Role of sympathetic activity in blood pressure reduction with low caloric regimen. *Hypertension 17*: 965–968.

Lane, M. A., Baer, D. J., Rumpler, W. V., Weindruch, R., Ingram, D. K., Tilmont, E. M., Cutler, R. G., & Roth, G. S. (1996). Calorie restriction lowers body temperature in rhesus monkeys, consistent with a postulated anti-aging mechanism in rodents. *Proc. Natl. Acad. Sci., USA 93*: 4159–4164.

Lane, M. A., Baer, D. J., Tilmont, E. M., Rumpler, W. V., Ingram, D. K., Roth, G. S., & Cutler, R. G. (1995). Energy balance in rhesus monkeys subjected to long-term dietary restriction. *J. Gerontol.: Biol. Sci. 50A*: B295–B302.

Lane, M. A., Black, A., Handy, A. M., Shapses, S. A., Tilmont, E. M., Kiefer, T. L., Ingram, D. K., & Roth, G. S. (2001). Energy restriction does not alter bone mineral metabolism or reproductive cycling and hormones in female rhesus monkeys. *J. Nutrition 131*: 820–827.

Lane, M. A., Ingram, D. K., Ball, S. S., & Roth, G. S. (1997a). Dehydroepiandrosterone sulfate: A biomarker of primate aging slowed by caloric restriction. *J. Clin. Endocrinol. Metab. 82*: 2093–2096.

Lane, M. A., Ingram, D. K., & Roth, G. S. (1997b). Beyond the rodent model: Calorie restriction in rhesus monkey. *Age 20*: 45–56.

Lass, A., Sohal, B. N., Weindruch, R., Forster, M., & Sohal, R. S. (1998). Caloric restriction prevents age-associated accrual of oxidative damage to mouse skeletal muscle mitochondria. *Free Radic. Biol. Med. 25*: 1089–1097.

Lee, J., Duan, W., Long, J. M., Ingram, D. K., & Mattson, M. P. (2000). Dietary restriction increases survival of newly-generated neural cells and induces BDNF expression in the dentate gyrus of rats. *J. Mol. Neurosci. 15*: 99–108.

Lee, Y. C. P., King, J. T., & Visscher, M. (1952). Influence of protein and calorie intake upon certain reproductive functions and carcinogenesis in the mouse. *Am. J. Physiol. 168*: 391–399.

Leung, F. C., Aylsworth, C. F., & Meites, J. (1983). Counteraction of underfeeding-induced inhibition of mammary tumor growth in rats by prolactin and estrogen administration. *Proc. Soc. Exp. Biol. Med. 173*: 159–163.

Levin, P., Janda, J. K., Joseph, J. A., Ingram, D. K., & Roth, G. S. (1981). Dietary restriction retards the age-associated loss of rat striatal dopaminergic receptors. *Science 214*: 561–562.

Lewis, S. E. M., Goldspink, D. F., Phillips, J. G., Merry, B. J., & Holehan, A. M. (1985). The effects of aging and chronic dietary restriction on whole-body growth and protein turnover in the rat. *Exp. Gerontol. 20*: 253–260.

Liepa, G. U., Masoro, E. J., Bertrand, H. A., & Yu, B. P. (1980). Food restriction as a modulator of age-related changes in serum lipids. *Am. J. Physiol. 238*: E253–E257.

Lintern-Moore, S. & Everitt, A. V. (1978). The effect of restricted food intake on the size and composition of the ovarian follicle population in the Wistar rat. *Biol. Reprod. 19*: 688–691.

London, E. D., Waller, S. B., Ellis, A. T., & Ingram, D. K. (1985). Effect of intermittent feeding on neurochemical markers in aging rat brain. *Neurobiol. Aging 6*: 199–204.

Luhtala, T. A., Roecker, E. R., Pugh, T., Feuers, R. J., & Weindruch, R. (1994). Dietary restriction attenuates age-related increases in rat skeletal muscle antioxidant enzyme activity. *J. Gerontol.: Biol. Sci. 49*: B231–B238.

Lynch, C. D., Cooney, P. T., Bennett, S. A., Thornton, P. L., Khan, A. S., Ingram, R. L., & Sonntag, W. E. (1999). Effect of moderate caloric restriction on cortical microvascular density and local cerebral blood flow in aged rats. *Neurobiol. Aging 20*: 191–200.

MacGibbon, M. F. (1996). Ageing as upregulation failure. *Med. Hypoth. 46*: 523–527.

MacGibbon, M. F., Walls, R. S., & Everitt, A. V. (2001). An age-related decline in melatonin secretion is not altered by food restriction. *J. Gerontol.: Biol. Sci. 56A*: B21–B26.

Magnuson, K. R. (1998). Aging of glutamate receptors: correlations between binding and spatial memory performance in mice. *Mech. Ageing Dev. 104*: 227–248.

Magnuson, K. R. (2001). Influence of diet restriction on NMDA receptor subunits and learning during aging. *Neurobiol. Aging 22*: 613–627.

Major, D. E., Kesslak, J. P., Cotman, C. W., & Finch, C. E. (1997). Life-long dietary restriction attenuates age-related increases in hippocampal glial fibrillary acidic protein mRNA. *Neurobiol. Aging 18*: 523–526.

Markowska, A. (1999). Life-long diet restriction failed to retard cognitive aging in Fischer-344 rats. *Neurobiol. Aging 20*: 177–189.

Markowska, A. & Breckler, S. J. (1999). Behavioral biomarkers of aging: Illustration of a multivariate approach for detecting age-related behavioral changes. *J. Gerontol.: Biol. Sci. 54A*: B549–B566.

Masoro, E. J. (2001). Dietary restriction: An experimental approach to the study of the biology of aging. In: E. J. Masoro (Ed.) *Handbook of the Biology of Aging*, 5th ed., (pp. 396–420). San Diego: Academic Press.

Masoro, E. J., Compton, C., Yu, B. P., & Bertrand, H. A. (1983). Temporal and compositional dietary restrictions modulate age-related changes in serum lipids. *J. Nutrition 113*: 880–892.

Masoro, E. J., Yu, B. P., & Bertrand, H. (1982). Action of food restriction in delaying the aging processes. *Proc. Natl. Acad. Sci. USA 79*: 4239–4241.

Mattison, J. A., Roth, G. S., Ingram, D. K., & Lane, M. A. (2001). Endocrine effects of dietary restriction and aging: The National Institute on Aging study. *J. Anti-Aging Med. 4*: 215–223.

Mattson, M. P., Duan, W., Lee, J., & Guo, Z. (2001). Suppression of brain aging and neurodegenerative disorders by dietary restriction and environmental enrichment: molecular mechanisms. *Mech. Ageing Dev. 122*: 757–778.

McCarter, R. J. M. (2000). Caloric restriction, exercise and aging. In: C. K. Sen, L. Parker, & O. Hanninen (Eds.) *Handbook of Oxidants and Antioxidants in Exercise* (pp. 797–829). Amsterdam: Elsevier Science.

McCarter, R. J. M. & McGee, J. R. (1989). Transient reduction in metabolic rate by food restriction. *Am. J. Physiol. 257*: E175–E179.

McCarter, R. J. M. & Palmer, J. (1992). Energy metabolism and aging: a lifelong study in Fischer 344 rats. *Am. J. Physiol. 263*: E448–E452.

McCarter, R. J. M., Masoro, E. J., & Yu, B. P. (1982). Rat muscle structure and metabolism in relation to age and food intake. *Am. J. Physiol. 242*: R89–R93.

McCarter, R. J. M., Shimokawa, I., Ikeno, Y., Higami, Y., Hubbard, G. B., Yu, B. P., & McMahan, C. A. (1997). Physical activity as a factor in the action of dietary restriction on aging: Effects in Fischer 344 rats. *Aging Clin. Exp. Res. 9*: 73–79.

McCay, C. M., Crowell, M. F., & Maynard, L. A. (1935). The effect of retarded growth upon the length of the life span and upon ultimate body size. *J. Nutrition 10*: 63–79.

McShane, T. M. & Wise, P. M. (1996). Life-long moderate caloric restriction prolongs reproductive life span in rats without interrupting estrous cyclicity: Effects on the gonadotropin-releasing hormone/luteinizing axis. *Biol. Reprod. 54*: 70–75.

McShane, T. M., Petersen, S. L., McCrone, S., & Keisler, D. H. (1993). Influence of food restriction on neuropeptide Y, propiomelanocortin, and luteinizing hormone-releasing hormone gene expression in sheep. *Biol. Reprod. 49*: 831–839.

Means, L. W., Higgins, J. L., & Fernandez, T. G. (1993). Mid-life onset of dietary restriction extends life and prolongs cognitive functioning. *Physiol. Behav. 54*: 503–508.

Merry, B. J. & Holehan, A. M. (1979). Onset of puberty and duration of fertility in rats fed a restricted diet. *J. Reprod. Fert. 57*: 253–259.

Merry, B. J. & Holehan, A. M. (1981). Serum profiles of LH, FSH, testosterone and 5α-DHT from 21 to 1000 days in *ad libitum*-fed and dietary restricted long-lived rats. *Exp. Gerontol. 16*: 431–444.

Merry, B. J. & Holehan, A. M. (1988). Dietary restriction and the neuroendocrinology of aging. In: A. V. Everitt, & J. R. Walton (Eds.), *Regulation of Neuroendocrine Aging* (pp. 61–73). Basel: Karger.

Merry, B. J., Lewis, S. E., & Goldspink, D. F. (1992). The influence of age and chronic restricted feeding on protein synthesis in small intestine of the rat. *Exp. Gerontol. 27*: 191–200.

Miller, R. A. (1991). Caloric restriction and immune function: developmental mechanisms. *Aging Clin. Exper. Res. 3*: 395–398.

Miller, R. A. & Harrison, D. E. (1985). Delayed reduction in T cell precursor frequencies accompanies diet-induced lifespan extension. *J. Immunol. 134*: 1426–1429.

Morgan, T. E., Rozovsky, I., Goldsmith, S. K., Stone, D. J., Yoshida, T., & Finch, C. E. (1997). Increased transcription of the astrocyte gene GFAP during middle-age is attenuated by food restriction: Implications for the role of oxidative stress. *Free Radic. Biol. Med. 23*: 524–528.

Morgan, T. E., Xie, Z., Goldsmith, S., Yoshida, T., Lanzrein, A. S., Stone, D., Rozonsky, I., Perry, G., Smith, M. A., & Finch, C. E. (1999). The mosaic of brain glial hyperactivity during normal aging and its attenuation by food restriction. *Neuroscience 89*: 687–699.

Moroi-Fetters, S. E., Mervis, R. F., London, E. D., & Ingram, D. K. (1989). Dietary restriction suppresses age-related changes in dendritic spines. *Neurobiol. Aging 10*: 317–322.

Morris, G. S., Surdyka, S. F., & Baldwin, K. M. (1990). Apparent influence of metabolism on cardiac isomyosin profile of food-restricted rats. *Am. J. Physiol. 258*: R346–R351.

Moscrip, T. D., Ingram, D. K., Lane, M. A., Roth, G. S., & Weed, J. L. (2000). Locomotor activity in female rhesus monkeys: Assessment of age and calorie restriction effects. *J. Gerontol.: Biol. Sci. 55A*: B373–B380.

Nelson, J. F. (1995). The potential role of selected endocrine systems in aging processes. In: E. J. Masoro (Ed.), *Handbook of Physiology,* Section 11, *Aging* (pp. 377–394). New York, Oxford University Press.

Nelson, J. F. & Felicio, L. S. (1985). Reproductive aging in the female: An etiologic perspective. *Rev. Biol. Res. Aging 2*: 251–314.

Nelson, J. F., Gosden, R. G., & Felicio, L. S. (1985). Effect of dietary restriction on estrous cyclicity and follicular reserves in aging C57BL/6J mice. *Biol. Reprod. 32*: 515–522.

Obin, M., Pike, A., Halbleib, M., Lipman, R., Taylor, A., & Bronson, R. (2000). Calorie restriction modulates age-dependent changes in the retinas of Brown Norway rats. *Mech. Ageing Dev. 114*: 133–147.

Orentreich, N., Brind, J. L., Vogelman, J. H., Andres, R., & Baldwin, H. (1992). Long-term longitudinal measurements of plasma dehydroepiandrosterone sulfate in normal men. *J. Clin. Endocrinol. Metab. 75*: 1002–1004.

Osborne, T. B. & Mendel, L. B. (1915). The resumption of growth after long continued failure to grow. *J. Biol. Chem. 23*: 439–454.

Oscai, L. B., Spirakis, C. N., Wolff, C. A., & Beck, R. J. (1972). Effect of exercise and food restriction on adipose tissue cellularity. *J. Lipid Res. 13*: 588–592.

O'Steen, W. K. & Landfield, P. W. (1991). Dietary restriction does not alter retinal aging in the Fischer 344 rat. *Neurobiol. Aging 12*: 455–462.

Pahlavani, M. A. (1998). Intervention in the aging of the immune system: Influence of dietary restriction, dehydroepiandrosterone, melatonin, and exercise. *Age 21*: 153–173.

Pahlavani, M. A., Cheung, H. T., Cai, N. S., & Richardson, A. (1990). Influence of dietary restriction and aging on gene expression in the immune system of rats. In: A. L. Goldstein (Ed), *Biomedical Advances in Aging* (pp. 259–270). New York: Plenum Publishing Corp.

Pahlavani, M. A., Harris, M. D., & Richardson, A. (1997). The increase in the induction of IL-2 expression with caloric restriction is correlated to changes in the transcription factor NFAT. *Cell. Immunol. 180*: 10–19.

Pedersen, W. A. & Mattson, M. P. (1999). No benefit of dietary restriction on disease onset or progression in amyotrophic lateral sclerosis in Cu/Zn-superoxide dismutase mutant mice. *Brain Res. 833*: 117–120.

Pitsikas, N. & Algeri, S. (1992). Deterioration of spatial and nonspatial reference and working memory in aged rats: protective effect of lifelong calorie restriction. *Neurobiol. Aging 13*: 369–373.

Pitsikas, N., Carli, M., Fidecka, S., & Algeri, S. (1990). Effect of life-long hypocaloric diet on age-related changes in motor and cognitive behavior in a rat population. *Neurobiol. Aging 11*: 417–423.

Quigley, K., Goya, R., & Meites, J. (1987). Rejuvenating effect of 10-week underfeeding period on estrous cycles in young and old rats. *Neurobiol. Aging 8*: 225–232.

Quigley, K., Goya, R., Nachreiner, R., & Meites, J. (1990). Effects of underfeeding and refeeding on GH and thyroid hormone secretion in young, middle-aged and old rats. *Exp. Gerontol. 25*: 447–457.

Ramsey, J. J., Colman, R. J., Binkley, N. C., Christensen, T. A., Gresl, T. A., Kemnitz, J. W., & Weindruch, R. (2000a). Dietary restriction and aging in rhesus monkeys: the University of Wisconsin study. *Exp. Gerontol. 35*: 1131–1149.

Ramsey, J. J., Harper, M-E., & Weindruch, R. (2000b). Restriction of energy intake, energy expenditure, and aging. *Free Radic. Biol. Med. 29*: 946–968.

Ramsey, J. J., Roecker, E. B., Weindruch, R., Baum, S. T., & Kemnitz, J. W. (1996). Thermogenesis of adult rhesus monkeys: Results through 66 months of dietary restriction. *FASEB J. 10*: A726.

Ramsey, J. J., Roecker, E. B., Weindruch, R., & Kemnitz, J. W. (1997). Energy expenditure of adult male rhesus monkeys during the first 30 mos. of dietary restriction. *Am. J. Physiol. 272*: E901–E907.

Reaven, G. M. & Reaven, E. P. (1981). Prevention of age-related hypertriglyceridemia by caloric restriction and exercise training. *Metabolism 30*: 982–986.

Reed, M. J., Penn, P. E., Li, Y., Birnbaum, R., Vernon, R. B., Johnson, T. S., Pendergrass, W. R., Sage, E. H., Abrass, I. B., & Wolf, N. S. (1996). Enhanced cell proliferation and biosynthesis mediate improved wound repair in refed, caloric-restricted mice. *Mech. Ageing Dev. 89*: 21–41.

Reiser, K., McGee, C., Rucker, R., & McDonald, R. (1995). Effects of aging and caloric restriction on extracellular matrix biosynthesis in a model of injury repair in rats. *J. Gerontol.: Biol. Sci. 50A*: B40–B47.

Ricketts, W. G., Birchenall-Sparks, M. C., Hardwick, J. P., & Richardson, A. (1985). Effect of age and dietary restriction on protein synthesis by isolated kidney cell. *J. Cell. Physiol. 125*: 492–498.

Riley-Roberts, M. L., Turner, R. I., Evans, P. M., & Merry, B. I. (1989). Failure of dietary restriction to influence natural killer activity in old rats. *Mech. Ageing Dev. 50*: 81–93.

Roecker, E. B., Kemnitz, J. W., Ershler, W. B., & Weindruch, R. (1996). Reduced immune responses in rhesus monkeys subjected to dietary restriction. *J. Gerontol.: Biol. Sci. 51A*: B276–B279.

Roth, G. S., Blackman, M. R., Ingram, D. K., Lane, M. A., Ball, S. S., & Cutler, R. G. (1993). Age-related changes in androgen levels in rhesus monkeys subjected to diet restriction. *Endocrine J. 1*: 227–234.

Roth, G. S., Ingram, D. K., & Joseph, J. A. (1984). Delayed loss of striatal dopaminergic receptors during aging of dietarily restricted rats. *Brain Res. 300*: 27–32.

Roth, G. S., Kowatch, M. A., Hengemihle, J., Ingram, D. K., Spangler, E. L., Johnson, L. K., & Lane, M. A. (1997). Effect of age and caloric restriction on cutaneous wound closure in rats and monkeys. *J. Gerontol.: Biol. Sci. 52A*: B98–B102.

Roth, G. S., Lesnikov, V., Lesnikov, M., Ingram, D. K., & Lane, M. L. (2001). Dietary caloric restriction prevents the age-related decline in plasma melatonin levels. *J. Clin. Endocrin. Metab. 86*: 3292–3295.

Sanderson, J. P., Binkley, N., Roecker, E. B., Champ, J. E., Pugh, T. D., Aspnes, L., & Weindruch, R. (1997). Influence of fat intake and caloric restriction on bone in aging male rats. *J. Gerontol.: Biol. Sci. 52A*: B20–B25.

Shimokawa, I. & Higami, Y. (1999). A role for leptin in the antiaging action of dietary restriction. *Aging: Clin. Exp. Res. 11: 380–382.*

Shimokawa, I. & Higami, Y. (2001). Leptin signaling and aging: insight from caloric restriction. *Mech. Ageing Dev. 122*: 1511–1519.

Snyder, D. L., Aloyo, V. J., Wang, W., & Roberts, J. (1998a). Influence of age and dietary restriction on norepinephrine uptake into cardiac synaptosomes. *J. Cardiovasc. Pharmacol. 32*: 896–901.

Snyder, D. L., Gayheart-Walsten, P. A., Rhie, S., Wang, W., & Roberts, J. (1998b). Effect of age, gender, rat strain, and dietary restriction, on norepinephrine release from cardiac synaptosomes. *J. Gerontol.: Biol. Sci. 53A*: B33–B41.

Spaulding, C. C., Walford, R. L., & Effros, R. B. (1997). The accumulation of non-replicative, non-functional, senescent T cells with age is avoided in calorically restricted mice by an enhancement of T-cell apoptosis. *Mech. Ageing Dev. 93*: 25–33.

Sprott, R. L. & Austad, S. N. (1996). Animal models for aging research. In: E. L. Schneider & J. W. Rowe (Eds.), *Handbook of the Biology of Aging,* 4th ed. (pp. 3–23). San Diego: Academic Press.

Stewart, J., Mitchell, J., & Kalant, N. (1989). The effect of life-long food restriction on spatial memory in young and aged Fischer 344 rats measured in the eight-arm radial and the Morris water mazes. *Neurobiol. Aging 10*: 669–675.

Stokkan, K.-A., Reiter, R. J., Nonaka, K. O., Lerchl, A., Yu, B. P., & Vaughn, M. K. (1991). Food restriction retards aging of the pineal gland. *Brain Res. 545*: 66–72.

Sun, D., Muthukumar, A. R., Lawrence, R. A., & Fernandes, G. (2001). Effects of calorie restriction on polymicrobial peritonitis induced by cecum ligation and puncture in young C57BL/6 mice. *Clin. Diag. Lab. Immunol. 8*: 1003–1008.

Swoap, S. J., Haddad, F., Bodell, P., & Baldwin, K. M, (1994). Effect of chronic energy deprivation in cardiac thyroid hormone receptor and myosin isoform expression. *Am. J. Physiol. 266*: E254–E260.

Tacconi, M. T., Lligona, L., Salmona, M., Pitsikas, N., & Algeri, S. (1991). Aging and food restriction: Effect on lipids of the cerebral cortex. *Neurobiol. Aging 12*: 55–59.

Taffet, G. E., Pham, T. T., & Hartley, C. J. (1997). The age-associated alterations in late diastolic function in mice are improved by caloric restriction. *J. Gerontol.: Biol. Sci. 52A*: B285–B290.

Tavernakis, N. & Driscoll, M. (2002). Caloric restriction and life span: a role for protein turnover? *Mech. Ageing Dev. 123*: 215–229.

Teillet, L., Ribiere, P., Gouraud, S., Bakola, J., & Corman, B. (2002). Cellular signaling, AGE accumulation, and gene expression in hepatocytes of lean aging rats fed ad libitum or food-restricted. *Mech. Ageing Dev. 123*: 427–439.

Thomas, J., Bertrand, H., Stacy, C., & Herlihy, J. T. (1993). Long-term caloric restriction improves baroeflex sensitivity in aging Fischer 344 rats. *J. Gerontol.: Biol. Sci. 48*: B151–B155.

Venkatraman, J. & Fernandes, G. (1992). Modulation of age-related alterations in membrane composition and receptor-associated immune functions by food restriction in Fischer 344 rats. *Mech. Ageing Dev. 63*: 27–44.

Venkatraman, J., Attwood, V. G., Turturro, A., Hart, R. W., & Fernandes, G. (1994). Maintenance of virgin T cells and immune functions by food restriction during aging in long-lived $B_6D_2F_1$ female mice. *Aging: Immunol Infect. Dis. 5: 13–26.*

Verdery, R. B. & Walford, R. L. (1998). Changes in plasma lipids and lipoproteins in humans during a 2-year period of dietary restriction in Biosphere-2. *Arch. Intern. Med. 158*: 900–906.

Verdery, R. B., Ingram, D. K., Roth, G. S., & Lane, M. A. (1997). Caloric restriction increases HDL_2 levels in rhesus monkeys *(Macaca mulatta)*. *Am. J. Physiol. 273*: E714–E719.

Vischer, M. B., King, J. T., & Lee, Y. C. P. (1952); Further studies on the influence of age and diet on reproductive senescence in strain A female mice. *Am. J. Physiol. 170*: 72–76.

Volk, M. J., Pugh, T. D., Kim, M. J., Frith, C. H., Dayne, R. A., Ershler, W. B., & Weindruch, R. (1994). Dietary restriction from middle age attenuates age-associated lymphoma development and interleukin 6 dysregulation in C57BL/6 mice. *Cancer Res. 54*: 3054–3061.

Walford, R. L. (1983). *Maximum Life Span*. New York: W. W. Norton.

Walford, R. L., Harris, S. B., & Gunion, M. W. (1992). The calorically restricted low-fat nutrient-dense diet in Biosphere 2 significantly lowers blood glucose, total leukocyte count, cholesterol, and blood pressure. *Proc. Natl. Acad., Sci. USA 89*: 11533–11537.

Ward, W. F. (1988a). Enhancement by food restriction of liver protein synthesis in aging Fischer 344 rats. *J. Gerontol.: Biol. Sci. 43*: B50–B53.

Ward, W. F. (1988b). Food restriction enhances the proteolytic capacity of the aging rat liver. *J. Gerontol.: Biol. Sci. 43*: B121–B124.

Weed, J. L., Lane, M. A., Roth, G. S., Speer, D. L., & Ingram, D. K. (1997). Activity measures in rhesus monkeys on long-term caloric restriction. *Physiol. Behav. 62*: 97–103.

Weindruch, R., Devens, B. H., Raff, H. V., & Walford, R. L. (1983). Influence of aging and diet restriction on natural killer cell activity in mice. *J. Immunol. 130*: 993–996.

Weindruch, R., Gotterman, S. R. S., & Walford, R. L. (1982a). Modification of age-related immune decline in mice dietarily restricted from or after adulthood. *Proc. Natl. Acad. Sci. USA 79*: 898–902.

Weindruch, R., Kristie, J. A., Cheney, K. E., & Walford, R. L. (1979). Influence of controlled dietary restriction on immunologic function and aging. *Fed. Proc. 38*: 2007–2016.

Weindruch, R., Kristie, J. A., Naeim, F., Muller, G. B., & Walford, R. L. (1982b). Influence of weaning-initiated dietary restriction on responses to cell mitogens and on splenic T cell levels in a long-lived F_1-hybrid mouse strain. *Exp. Gerontol. 17*: 49–61.

Weindruch, R., Lane, M. A., Ingram, D. K., Ershler, W. B., & Roth, G. S. (1997). Dietary restriction in rhesus monkeys: Lymphopenia and reduced mitogen-induced proliferation in peripheral blood mononuclear cells. *Aging Clin. Exp. Res. 9*: 304–308.

Yu, B. P., Masoro, E. J., & McMahan, C. A. (1985). Nutritional influences on aging of Fischer 344 rats: I. Physical, metabolic, and longevity characteristics. *J. Gerontol. 40*: 657–670.

Yu, B. P., Masoro, E. J., Murata, I., Bertrand, H. A., & Lynd, F. T. (1982). Life span study of SPF Fischer 344 male rats fed *ad libitum* or restricted diets: Longevity, growth, lean body mass, and disease. *J. Gerontol. 37*: 130–141.

Yu, B. P., Wong, G., Lee, H., Bertrand, H., & Masoro, E. J. (1984). Age changes in hepatic metabolic characteristics and their modulation by dietary manipulation. *Mech. Ageing Dev. 24*: 67–81.

Yu, Z. F., & Mattson, M. P. (1999). Dietary restriction and 2-deoxyglucose administration reduce focal ischemic brain damage and improve behavioral outcome: evidence for a preconditioning mechanism. *J. Neurosci. Res. 57*: 830–859.

Zhu, H., Guo, Q., & Mattson, M. P. (1999). Dietary restriction protects hippocampal neurons against death-promoting action of presenilin-1 mutation. *Brain Res. 842*: 224–229.

CHAPTER 5

Age-associated diseases

Contents

Since the late 1930s, many studies on a variety of mouse and rat strains have shown that CR delays the onset and/or slows the progression of many different age-associated diseases. Such diseases include neoplastic diseases, degenerative diseases, and autoimmune diseases (Bronson & Lipman, 1991; Hart et al., 1999; Roe et al., 1995; Turturro et al., 1994). Although CR retards the appearance and/or progression of most age-associated pathologic lesions in these rodent species, it must be noted that not all such lesions are so affected; e.g. it does not influence lung adenoma in female B6C3F1 mice (Lipman et al., 1999a) or testicular degeneration and pancreatic atrophy in rats (Lipman et al., 1999b).

Cancer

Although cancer can occur at all ages, the incidence of most cancers increases with increasing age in humans and animals (Dix, 1989). It has long been known that rats and mice on a long-term CR regimen show a markedly lower lifetime incidence of most types of tumors and/or a delay in the age of occurrence (Albanes, 1987; Berg & Simms, 1960;

RESEARCH PROFILES IN AGING
VOLUME 1 ISSN 1567-7184

Table 5-1
CR and the prevalence of tumors in male F-344 rats

Age range, months	AL rats with tumors			CR rats with tumors		
	Examined	Testes	Other	Examined	Testes	Other
2–11	12	0	0	14	0	0
12–17	17	3	3	14	0	0
18–23	45	40	19	11	1	1
24–29	46	43	24	33	27	10
30–35	1	1	1	35	24	21
36–43				17	14	14

Note: No AL (*ad libitum*-fed) rats were alive in the 36–43 months age range; the testicular tumors were Leydig cell tumors; the rats were either sacrificed or died spontaneously. (Data in table are from Yu et al. (1982) and Maeda et al. (1985).)

Conybeare, 1980; Huseby et al., 1945; Morris, 1945; Ross, 1959; Rous, 1914; Rusch, 1944; Saxton et al., 1944; Silberberg & Silberberg, 1955; Tannenbaum, 1940, 1942, 1945; Tannenbaum & Silverstone, 1953a; White, 1961). The findings of Maeda et al. (1985) and Yu et al. (1982) on the effects of age and CR started at 6 weeks of age on tumor prevalence in male F344 rats are summarized in Table 5-1. In addition, Ross and Bras (1971) found that CR for only 7 weeks immediately after weaning substantially reduces cancer risk during the lifetime of the rat, although much less effectively than CR throughout life. In contrast, CR initiated in rats at 6 months of age (young adults) is as effective as that initiated at 6 weeks of age (2 weeks post-weaning) in delaying the occurrence of cancers (Maeda et al., 1985). CR initiated at 1 year of age in B10C3F$_1$ mice inhibits the occurrence of lymphoma, but the effect is small and statistically marginal (Weindruch & Walford, 1982). Although CR started in middle age decreases lymphoma development and increases the age at death of tumor-bearing C57BL/6 mice, it also increases the percentage with cancer at the time of death (Pugh et al., 1999; Volk et al., 1994). Initiating CR at 18 or 26 months of age does not affect tumor burden in F344×BN F1 hybrid rats (Lipman et al., 1998). Total energy intake has been shown to be an important determinant of carcinogenesis in both mice and rats (Albanes, 1987; Ross et al., 1983).

Many mechanisms were proposed by Weindruch (1989) for the effect of CR on cancer; these include: diminished cellular oxidative damage; retardation of the age-associated decrease in immune function; modulation of hormonal and growth factors; reduced exposure to dietary carcinogens; reduced energy for tumor growth; less activation of carcinogens; and greater maintenance of DNA repair. It is surprising that CR delays spontaneous carcinogenesis in mice carrying the null mutation in the p53 tumor suppressor gene (Hursting et al., 1994). Also, surprisingly, an elevated environmental temperature, which prevents daily periods of hypothermia in C57BL/6 mice on a CR regimen, decreases the efficacy of this dietary regimen in retarding the development of lymphoma in this strain (Koizumi et al., 1996). It has been suggested that CR retards the occurrence of cancers by decreasing cell division (Rogers et al., 1993) and by increasing apoptosis (James & Muskhelishvili, 1994; Warner et al., 1995). Indeed, Dunn et al. (1997) believe that CR slows the progression of bladder cancer in mice by decreasing IGF-1 levels, thereby promoting apoptosis over cell proliferation. CR also significantly influences the hepatic enzymes involved in the metabolism of carcinogens in rodents (Manjgaladze et al., 1993).

When female B10C3F$_1$ mice are maintained on CR from a young age, they have fewer cancers at the time of spontaneous death than those fed *ad libitum* (Cheney et al., 1983). In contrast, long-term CR delays the occurrence of cancers in male F344 rats (Shimokawa et al., 1991), but at the time of spontaneous death, the fraction of rats with cancer is not less than that of rats fed *ad libitum* (Maeda et al., 1985). The reason for this apparent paradox is that nonneoplastic age-associated diseases underlie the death of many *ad libitum*-fed rats at a relatively young age (Shimokawa et al., 1991). Neoplastic disease was a major contributor to the spontaneous death of 48% of male F344 rats fed casein-containing semi-synthetic diet *ad libitum* and 60% of rats fed a similar diet but restricted to 60% of the *ad libitum* intake from 6 weeks of age on (Shimokawa et al., 1993a). Keenan et al. (1995b) reported that in Sprague-Dawley rats studied until 106 weeks of age, there was no difference at 106 weeks of age in overall tumor incidence between those fed *ad libitum* and those on a CR diet, even though CR delayed the onset of tumors. However, Christian et al. (1998) found that CR reduced the prevalence of endocrine-mediated tumors in 104-week-old Sprague-Dawley rats.

Saxton et al. (1944) reported a delay in onset of spontaneous leukemia in mice on a long-term CR regimen. It also delays the onset and decreases the incidence of lymphoma in C57BL/6 and B6C3F1 mice (Blackwell et al., 1995; Cheney et al., 1980; Sheldon et al., 1995a). While CR delays the occurrence of leukemia in F344 rats, it does not retard its progression, i.e., the time between onset and death (Shimokawa et al., 1993b). The duration of CR is an important determinant of the delayed occurrence of leukemia in rats (Higami et al., 1994). However, because of increased longevity of F344 rats on a CR regimen, they have a higher lifetime incidence of leukemia than rats fed *ad libitum* (Thurman et al., 1994).

CR delays the onset and reduces the lifetime incidence of mammary cancer in mice (Fernandes et al., 1976; Tannenbaum, 1940; Vischer et al., 1942; White et al., 1944) and also retards its growth (Tannenbaum & Silverstone, 1953b). Sarkar et al. (1982) suggested that CR slows mammary cancer growth, in part, by reducing the blood level of prolactin. It also has been suggested that CR's inhibition of mammary tumorigenesis in CH3/Ou mice results from a decrease in serum prolactin and a reduction of mammary epithelial kinetics (Engelman et al., 1991, 1993); the authors proposed that by acting together, these suppress mouse mammary tumor virus transcription and minimize the risk of activating protooncogenes. Indeed, Koizumi et al. (1990) found that CR decreases the level of mouse mammary tumor virus mRNA in the mammary glands of mice. Fernandes et al. (1995) reported that CR decreases mammary tumor incidence in mouse mammary tumor virus/v-Ha-ras transgenic (Onco) mice; they suggest that it does so by altering the expression of cytokines, oncogenes, and free-radical scavenging enzymes.

CR also inhibits mammary carcinogenesis in rats. A reduction in DNA synthesis by rat mammary tissues has been suggested as the underlying anticarcinogenic mechanism (Sinhu et al., 1988). In Sprague-Dawley female rats, CR delays the onset of mammary cancer but not its progression (Keenan et al., 1995b). In female F344 rats, CR decreases the lifetime incidence of mammary cancer (Thurman et al., 1994).

CR decreases the occurrence of pituitary tumors in Wistar rats (Tucker, 1979). It delays the onset of pituitary tumors in Sprague-Dawley rats, but not their progression (Keenan et al., 1995b). In F344 rats, CR delays the onset, slows the progression (Higami et al., 1995), and decreases the lifetime incidence of pituitary tumors (Thurman et al., 1994). Pituitary tumors are also markedly reduced by CR in female C57BL/6 mice (Blackwell et al., 1995).

Initiated at 6 weeks of age, CR delays the occurrence of pheochromo-cytoma in male F344 rats and slows its progression (Higami et al., 1995). The prevalence of pheochromocytoma is reduced in F344 × BN F_1 hybrid rats when CR is initiated as late as 73 weeks of age (Lipman et al., 1998).

CR delays the occurrence of interstitial cell tumors in the testes of male F344 rats (Higami et al., 1995; Maeda et al., 1985; Thurman et al., 1994) and male F344×BN F_1 hybrid rats (Thurman et al., 1995). Also, CR reduces the expression of testicular cytochrome P450 enzymes involved in the activation of carcinogens, a factor that may play an important role in the delayed appearance of these tumors (Seng et al., 1996).

In addition, CR delays the occurrence of pancreatic islet tumors in male F344 rats, but not their progression (Higami et al., 1995), and it decreases the occurrence of thyroid tumors in female C57BL/6 mice (Blackwell et al., 1995). It also reduces the incidence and delays the onset of metastatic prostate adenocarcinomas and hepatomas in Lobund-Wistar rats (Pollard et al., 1989).

The incidence of skin tumors in Wistar rats is decreased by CR (Tucker, 1979). It also delays the onset and reduces the incidence of oral squamous cell carcinomas in BN rats (Thurman et al., 1997).

CR decreases the onset of liver neoplasia in mice (Blackwell et al., 1995; Tucker, 1979), and there is evidence that it may do so by enhancing the elimination of preneoplastic hepatocytes by apoptosis (Muskhelishvili et al., 1996). It also decreases the occurrence of lung tumors in mice (Koizumi et al., 1993; Larsen & Heston, 1945; Tannenbaum, 1940).

In addition to retarding spontaneously occurring tumors, more than 50 years ago it was reported that CR inhibits the induction of neoplasia by chemical carcinogens (Lavik & Bowman, 1943; Tannenbaum, 1944). A number of investigators have reported that the induction of mammary tumors by 7,12-dimethylbenz(*a*)anthracene in rats is decreased by CR (Heuson & Legros, 1972; Klurfeld et al., 1989b; Kritchevsky et al., 1984; Sylvester et al., 1982, 1984). This action of CR occurs during the promotion period (Klurfeld et al., 1987, 1989a). Boissonnealt et al. (1986) suggested that this effect relates to a complex interaction of energy intake, energy retention, and body size.

The induction of colon tumors by 1,2,-dimethylhydrazine in rats is inhibited by CR (Klurfeld et al., 1987; Kritchevsky et al., 1986). This inhibition occurs during the promotion period (Klurfeld et al., 1987, 1989a). CR also decreases the induction of colon tumors by azomethane in rats (Kumar et al., 1990; Reddy et al., 1987). CR reduces methylazomethanol acetate-induced tumors in the small intestine of rats, but it has no such effect on *N*-methylnitrosourea-induced tumors of the colon (Pollard et al., 1984; Pollard & Luckert, 1985).

Grasl-Kraup et al. (1994) reported that CR reduced by 50% nafenopin-induced hepatocellular adenomas and carcinomas in rats; they believe CR does so by eliminating preneoplastic cells through apoptosis. CR inhibits azaserine-induced pancreatic exocrine neoplasms in male Wistar/Lewis rats (Roebuck et al., 1981a); to do so, the restriction must include the initiation period (Roebuck et al., 1981b).

CR suppresses the occurrence of asbestos-induced lung tumors in A/J mice (Koizumi et al., 1993). Pashko and Schwartz (1992) found that the promotion of skin papilloma by 12-*O*-tetradecanoylphorbol-13-acetate in mice is suppressed by CR. Skin tumors induced by ultraviolet irradiation are also reduced by CR (Kritchevsky & Klurfeld, 1986).

CR decreases the occurrence of tumors in rats exposed to ionizing radiation (Gross & Dreyfus, 1984). It also inhibits the development of leukemia in mice following exposure to ionizing radiation (Gross & Dreyfus, 1986).

Lipman (2002) studied the effects of CR on the mortality characteristics of eight strains of mice (A/J, BALB/c, C3H, C57BL/6, DBA/2J, FVB/J, NMRI, and 129/J), which were treated at 9 weeks of age with 7,12-dimethyl[a]anthracene. CR increased the average age at death in all strains except FVB/J. However, the magnitude of its effect varied among the strains, ranging from an increase of 24.9–63.5%.

Kidney disease

As human life expectancy has increased, the lifetime incidence of end-stage renal disease has increased markedly (Stern et al., 2001). Indeed, it has become a major problem in geriatric medicine.

End-stage nephropathy is also commonly seen in old rats (Anver & Cohen, 1979). Renal lesions begin in young adulthood and progress in severity with age, often ultimately culminating in renal failure. Studies using a variety of rat strains have shown that CR, begun at weaning or soon after, markedly delays the onset of these lesions and slows the progression in severity (Berg & Simms, 1960; Bras & Ross, 1964; Davis et al., 1983; Everitt et al., 1982; Nolen, 1972; Saxton & Kimball, 1941; Tucker et al., 1976; Yu et al., 1982). CR also retards the development of kidney pathology in rhesus monkeys (Hansen et al., 1999).

Maeda et al. (1985) found that more than 50% of male F344 rats, fed a standard semi-synthetic diet *ad libitum*, had end-stage chronic nephropathy at the time of spontaneous death (Table 5-2). However, less than 2% of the rats on CR since 6 weeks of age exhibited end-stage nephropathy at the time of spontaneous death (Table 5-2), even though the rats on CR died at much older ages. Moreover, when CR was started at 6 months of age (young adults), it was as effective in retarding chronic nephropathy as when started at 6 weeks of age. On the other hand, when CR was only maintained from 6 weeks of age to 6 months of age, it had little effect on this disease process. Many of the *ad libitum*-fed rats exhibited renal hyperparathyroid disease (a complex of lesions

Table 5-2
CR and chronic nephopathy in male F344 rats dying spontaneously

Dietary regimen	Median life span, days	% of rats with nephropathy of grade					
		0	1	2	3	4	E
AL	701	0	1	17	10	21	51
CR	1057	11	62	20	5	0	2

Note: AL denotes *ad libitum*-fed; Nephropathy Grading: 0 denotes no lesions, 1–4 denotes lesions of progressively increasing severity; E denotes end-stage lesions; Grade 4 lesions were associated with moderately elevated serum creatinine and Grade E lesions with markedly elevated levels. (Data from Maeda et al., 1985.)

consisting of nephropathy, parathyroid hyperplasia, osteodystrophy, and metastatic calcifications); none of the rats on long-term CR showed this complex of lesions. CR effectively retards chronic nephropathy in male F344 rats fed a diet containing 35% casein (Masoro et al., 1989); the protein intake of the rats on this CR diet was the same as that of the rats fed the standard diet *ad libitum* and 1.7 times greater on a per unit body mass basis. This study shows that CR protects the kidneys from the well-known, long-term damaging action of high-protein diets.

Gumprecht et al. (1993) found that CR decreases the development of chronic nephropathy in Sprague-Dawley rats, and that this correlates with the mitigation of the early occurrence of glomerular hypertrophy. Keenan et al. (1995a) proposed that hyperfiltration causes glomerular damage in *ad libitum*-fed rats, leading to proliferation of mesangial, epithelial, and endothelial glomerular cells. This is followed by glomerular hypertrophy with continued accumulation of mesangial matrix, thickening of the glomerular basement membrane, and endothelial cell damage; ultimately, glomerular sclerosis occurs, and there is loss of functional nephrons. These findings suggest that CR inhibits the occurrence and progression of chronic nephropathy by retarding the early development of glomerular hypertrophy.

Cardiovascular disease

Cardiomyopathy is common in most strains of rats, and it increases in incidence and severity with increasing age (Lewis, 1992). In male F344 rats, CR decreases both the incidence and age-associated increase in severity of this cardiac pathology (Maeda et al., 1985); Table 5-3

Table 5-3
CR and cardiomyopathy in male F344 rats dying spontaneously

Dietary regimen	Median life span, days	% of rats with cardiomyopathy of grade			
		0	1	2	3
AL	701	11	27	41	21
CR	1057	25	50	25	0

Note: AL denotes *ad libitum*-fed; Cardiomyopathy Grading: 0 denotes no lesions, Grades 1–3 denote lesions of progressively increasing severity. (Data from Maeda et al., 1985.)

summarizes the findings on rats that died spontaneously. In a study carried out in our laboratory (Shimokawa et al., 1993a), cardiomyopathy was a major factor in the spontaneous death of male F344 rats, contributing to the death of 36% of those fed *ad libitum* and 11% of those on a CR regimen. Moreover, it should be noted that most of the CR rats were considerably older at the time of spontaneous death than those fed *ad libitum*. CR has also been found to partially protect rats of the Sprague-Dawley and Lobund-Wistar strains from age-associated occurrence and progression of cardiomyopathy (Cornwell et al., 1991; Keenan et al., 1995a; Snyder et al., 1990).

In several studies with Spontaneously Hypertensive Strain (SHR) rats, CR significantly lowered the arterial blood pressure (Fernandes et al., 1986; Notargiacomo & Fries, 1981; Overton et al., 1997: Wright et al., 1981). However, in two other studies with the SHR rat strain, CR had no effect on arterial blood pressure (Gradin & Persson, 1990; Susic et al., 1990). As discussed in Chapter 4, CR has been shown to decrease arterial blood pressure in normotensive rat strains, but the magnitude of this action is small (Swoap et al., 1995). In a lifetime study of male F344 rats, CR had no effect on the systolic blood pressure (Yu et al., 1985). Lloyd (1984) reported that CR dramatically reduces the incidence of hypertension-associated pathologic lesions in SHR rats. CR also decreases the arterial blood pressure of the hypertensive stroke-prone (SHR-SP) strain of rats (Stevens et al., 1998).

Coronary heart disease and stroke are common age-associated cardiovascular diseases in humans, and atherosclerosis is a major pathological lesion underlying these diseases. Koletsky and Puterman (1977) studied a genetically obese strain of rats that exhibit hypertriglyceridemia, hypercholesterolemia, and atherosclerotic lesions; they found that CR eliminates both hypertriglyceridemia and hypercholesterolemia and decreases atherosclerotic lesions in this strain. Here, it must be pointed out that most rat strains are not good models for the study of atherosclerosis because they do not exhibit the age-associated occurrence of this lesion. However, the rhesus monkey does suffer age-associated atherosclerosis and is therefore widely used for the study of atherogenesis. Although the influence of CR on atherosclerosis has not been directly measured in the rhesus monkey, it does have effects that should reduce the risk of developing this lesion (Lane et al., 1998). Verdery et al. (1997) reported that CR (30% restriction for 6–7 years) in adult rhesus monkeys increases the plasma levels of the HDL_{2b} sub-fraction of high-density lipoproteins (HDL), the sub-fraction associated with protection from atherosclerosis. Edwards et al. (1998) studied the binding of plasma low-density lipoproteins (LDL) of rhesus monkeys to proteoglycans isolated from the arterial wall; CR decreased this binding, which led the authors to suggest that CR may decrease the retention of LDL by the arterial wall, thereby reducing their atherogenic potential. In addition, CR lowers the arterial blood pressure in rhesus monkeys (Lane et al., 1999), thus decreasing another risk factor. Findings on the human participants of the Biosphere 2 study are in accord with this view (Verdery & Walford, 1998; Walford et al., 1992, 2002). The participants underwent CR for a period of two years; during this time, they exhibited a decrease in both arterial blood pressure and plasma cholesterol level, functional changes that decrease the risk of atherosclerotic disease.

Diabetes

Type II diabetes (noninsulin-dependent diabetes mellitus) is a major age-associated disease in humans. The manner in which it acts on the insulin–carbohydrate metabolism system of rodents and nonhuman primates indicates that CR should function to decrease the risk of developing Type II diabetes (Lane et al., 1998; Masoro, 2001). The many effects of CR on the insulin–carbohydrate metabolism system will be presented in detail in Chapter 6. Meanwhile, this chapter will focus on the evidence that CR does, indeed, forestall the occurrence of Type II diabetes in rhesus monkeys.

Hansen and Bodkin (1993) maintained eight adult male rhesus monkeys at a stabilized weight for 5–9 years by restricting their food intake and allowed another 19 monkeys to eat *ad libitum*. When the monkeys were approximately 18 years of age, four of the *ad libitum*-fed were overtly diabetic and six others had a significantly reduced glucose tolerance. However, none of the monkeys on the restricted diet developed these problems. This study provides strong evidence that CR does delay and may even totally prevent the occurrence of Type II diabetes in the rhesus monkey.

Neurodegenerative disease

In humans, age-associated neurodegenerative diseases, such as Alzheimer's disease and Parkinson's disease, are a major geriatric problem. Similar neurological disorders can be induced in rats and mice by pharmacological, genetic, and other experimental manipulations and, as discussed in Chapter 4, CR suppresses the development of many of these experimentally induced lesions. A review article by Mattson et al. (2001) provides an in-depth discussion of these rat and mouse studies.

Immune disease

Because of their increased incidence in old age, certain human auto-immune diseases (e.g. rheumatoid arthritis and thyroiditis) are a particular problem for the elderly. CR has been shown to suppress autoimmune diseases in susceptible strains of mice.

The B/W mouse strain is susceptible to an autoimmune disease that leads to renal failure and death at an early age; the disease in this mouse model closely resembles human systemic lupus erythematosus. When started at an early age, CR more than doubles the length of life of these mice (Fernandes et al., 1976). Indeed, when started as late as 4–5 months of age, CR retards the development of nephritis in B/W mice (Friend et al., 1979). It also slows the age-associated increase in circulating immune complexes (Safai-Kutti et al., 1980) and markedly decreases the deposition of gamma-globulin in the glomerular capillaries of these mice (Fernandes et al., 1978a). With increasing age, CR preserves the B/W mouse's production of and response to interleukin-2 (Jung et al., 1982). A reduced production of proinflammatory cytokines by the kidneys and peripheral blood T-lymphocytes is associated with the delay in onset of autoimmune renal disease (Chandrasekar & Fernandes, 1994; Chandrasekar et al., 1995b; Jolly & Fernandes, 1999). CR also blunts the age-associated alterations in the proportions and functions (cytokine and immunoglobulin secretion) of T- and B-lymphocytes in mesenteric lymph nodes of B/W mice (Lim et al., 2000). The extension of life is paralleled by the prevention of age-associated increases in interferon-γ (INF-γ), transforming growth factor-β, interleukin-12 (IL-12), and interleukin-10 (IL-10) at both protein and mRNA levels as well in the localization of nuclear factor κB in kidney nuclei (Jolly et al., 2001). In addition, CR also increases the activity of antioxidant enzymes in the kidneys of young B/W mice (Jolly et al., 2001). It also decreases renal platelet-derived growth factor A and thrombin receptor expression (Troyer et al., 1997), reduces the expression of serum gp70 and immune complex deposition (Fernandes et al., 1978a; Izui et al., 1981), and decreases glomerular expression of plasminogen activator inhibitor Type 1 (Troyer et al., 1995).

Female B/W mice are known to develop salivary gland lesions similar to those seen in human Sjogren's syndrome. CR decreases the severity of these lesions, and this amelioration is associated with increased expression of transforming growth factor-β1 and decreased expression of proinflammatory cytokines (Chandrasekar et al., 1995a). CR also lowers the mRNA expression of INF-γ and IL-10 in the submandibular salivary gland and the secretion of polymeric immunoglobulin by this gland (Muthukumar et al., 2000).

The BXSB mouse strain develops a lupus-like disease associated with B-cell hyperplasia in the peripheral lymphoid organs. Male mice develop inflammatory vascular disease of the heart and an autoimmune renal disease that differs somewhat from that seen with the B/W strain. CR increases the median length of life of BXSB mice and suppresses the development of autoimmune disease (Kubo et al., 1992). It also inhibits splenomegaly and the decline in IL-2 production.

The *kdkd* mouse strain is also autoimmune-prone and suffers from renal disease; however, its renal disease is histologically very different from that of the B/W strain. CR inhibits the autoimmunity directed toward erythrocytes in *kdkd* mice, retards the development of renal disease, and increases length of life (Fernandes et al., 1978b).

CR also prolongs the life of the MRL/Mp-lpr/lpr strain of mice (Kubo et al., 1984); this strain suffers from autoimmune lymphoproliferative disease. CR prevents the massive lymphoadenopathy and the splenomegaly as well as histological abnormalities in the thymus, spleen, lymph nodes and kidneys of these mice (Fernandes & Good, 1984). In addition to blunting the age-associated increase in "double-negative" T-cells, it maintains responsiveness of lymphocytes to mitogens and high levels of dexamethasone-induced apoptosis (Luan et al., 1995).

Ogura et al. (1989a) found that three mouse strains (NZB, MRL/lpr, and BXSB) respond to CR by decreasing the proliferation rate of lymphoid cells in the thymus, spleen, and mesenteric lymph nodes. Based on this commonality, the investigators suggest that decreased proliferation plays a role in CR's ability to inhibit autoimmune diseases. They also found that in four strains of autoimmune-prone mice (NZB, B/W, MRL/lpr, and BXSB), CR decreases the absolute and relative numbers of

Ly-1$^+$ B-lymphocytes to levels similar to those of long-lived mouse strains fed *ad libitum* (Ogura et al., 1989b).

Cataracts and glaucoma

A common age-associated problem for humans is the development of cataracts. The Emory strain of mice has been widely used as an animal model for the study of this human condition because this strain gradually develops lens cataracts with increasing age, with full lens opacification occurring by late life in most of the mice. CR retards the development of cataracts in this mouse strain (Taylor et al., 1989, 1995). With a 40% reduction in caloric intake below that of *ad libitum*-fed mice, 2% of the mice at 11 months of age exhibit bilateral grade 5 cataracts compared to 17% of those fed *ad libitum* (Mura et al., 1993). At 22 months of age, 18% of the CR mice exhibit bilateral grade 5 cataracts, compared to 90% of those fed *ad libitum* (Mura et al., 1993). Although the development of cataracts in this mouse strain is thought to stem from accumulated lens oxidative insult, evidence to date argues against attributing CR's anti-cataract action to an enhancement of antioxidant defenses (Gong et al., 1997).

Wolf et al. (2000) studied the development of cataracts in two rat strains (BN and F344) and three mouse strains (C57BL/6, C57BL/6×DBA/2 F$_1$, and C57BL/6 × C3H F1). They found that cataracts develop with advancing age in all five strains, and that CR delays the onset and slows progression in severity of cataracts in all the strains except F344 rats. It should be noted that of these five strains, only the F344 rats are albinos.

Glaucoma is another common age-associated human eye problem. The DBA/2 strain of mouse develops this problem, and CR has been found to retard its occurrence and slow its progression (Sheldon et al., 1995b).

Osteoarthritis

Mice undergo an age-associated degeneration of the vertebrae as a result of osteoarthritis, and CR ameliorates this disease process (Sheldon et al., 1996). CR also retards the occurrence of osteoarthritis in dogs. Kealy et al. (1997) maintained two groups of Labrador Retrievers, one allowed to eat *ad libitum* and the other given 75% of the food intake of the *ad libitum*-fed group. CR decreased the frequency and severity of osteoarthritis in these dogs.

References

Albanes, D. (1987). Total calories, body weight, and tumor incidence in mice. *Cancer Res. 47*: 1987–1992.

Anver, M. R. & Cohen, B. J. (1979). Lesions associated with aging. In: H. J. Baker, J. R. Lindsay, & S. H. Weisbroth, (Eds.), *The Laboratory Rat*, Vol. 8 (pp. 378–399). New York: Academic Press.

Berg, B. N. & Simms, H. S. (1960). Nutrition and longevity in the rat. II. Longevity and onset of disease with different levels of food intake. *J. Nutrition 71*: 255–263.

Blackwell, B-N, Bucci, T. J., Hart, R. W., & Turturro, A. (1995). Longevity, body weight, and neoplasia in *ad libitum*-fed and diet-restricted C57BL/6 mice fed NIH-31 open formula diet. *Toxicol. Path. 23*: 570–590.

Boissonneault, G. A., Elson, C. E., & Pariza, M. W. (1986). Net energy effects of dietary fat on chemically induced mammary carcinogenesis in F344 rats. *J. Natl. Cancer Inst. 76*: 335–338.

Bras, G. & Ross, M. H. (1964). Kidney disease and nutrition in the rats. *Toxicol. Pharmacol. 6*: 246–262.

Bronson, R. T. & Lipman, R. D. (1991). Reduction in rate of occurrence of age related lesions in dietary restricted laboratory mice. *Growth Dev. Aging 55*: 169–184.

Chandrasekar, B. & Fernandes, G. (1994). Decreased proinflammatory cytokines and increased antioxidant gene expression by n-3 lipids in murine lupus nephritis. *Biochem. Biophys. Res. Commun. 200*: 893–898.

Chandrasekar, B., McGriff, H. S., Aufdermorte, T. B., Troyer, D., Talal, N., & Fernandes, G. (1995a). Effects of calorie restriction on transforming growth factor $\beta1$ and proinflammatory cytokines in murine Sjogren's syndrome. *Clin. Immunol. Immunopath. 76*: 291–296.

Chandrasekar, B., Troyer, D. A., Venkatraman, J. T., & Fernandes, G. (1995b). Dietary omega-3 fatty acids delay the onset and progression of autoimmune lupus nephritis by inhibiting transforming growth factor β mRNA and protein expression. *J. Autoimmun. 8*: 381–393.

Cheney, K. E., Liu, R. K., Smith, G. S., Leung, R. E., Mickey, M. R., & Walford, R. L. (1980). Survival and disease patterns in C57/BL/6 mice subjected to undernutrition. *Exp. Gerontol. 15*: 237–258.

Cheney, K. E., Liu, R. K., Smith, G. S., Meredith, P. J., Mickey, M. R., & Walford, R. L. (1983). The effect of dietary restriction of varying duration on survival, tumor patterns, immune function, and body temperature in B10C3F₁ female mice. *J. Gerontol. 38*: 420–430.

Christian, M. S., Hoberman, A. M., Johnson, M. D., Brown, W. R., & Bucci, T. J. (1998). Effect of dietary optimization on growth, survival, tumor incidence, and clinical chemistry parameters in CD Sprague-Dawley and Fischer 344 rats: a 104 week study. *Drug Chem. Toxicol. 21*: 97–117.

Conybeare, G. (1980). Effect of quality and quantity of diet on survival and tumour incidence in outbred Swiss mice. *Food Cosmet. Toxicol. 18*: 65–75.

Cornwell, G. G., Thomas, B. P., & Snyder, D. L. (1991). Myocardial fibrosis in aging germ-free and conventional Lobund-Wistar rats: The protective effect of diet restriction. *J. Gerontol.: Biol. Sci. 46*: B167–B169.

Davis, T. A., Bales, C. W., & Beauchene, R. E. (1983). Differential effects of dietary restriction and protein restriction in the aging rat. *Exp. Gerontol. 18*: 427–435.

Dix, D. (1989). The role of aging in cancer incidence: an epidemiological study. *J. Gerontol. 44*: (Special Issue) 10–18.

Dunn, S. E., Kari, F. W., French, J., Leininger, J. R., Travlos, G., Wilson, R., & Barrett, J. C. (1997). Dietary restriction reduces insulin-like growth factor 1 levels, which modulates apoptosis, cell proliferation, and tumor progression in p^{53}-deficient mice. *Cancer Res. 57*: 4667–4672.

Edwards, I. J., Rudel, L. L., Terry, J. G., Kemnitz, J. W., Weindruch, R., & Cefalu, W. T. (1998). Caloric restriction in rhesus monkeys reduces low density lipoprotein interaction with arterial proteoglycans. *J. Gerontol.: Biol. Sci. 53A*: B443–B448.

Engelman, R. W., Day, N. K., & Good, R. A. (1993). Calories, parity, and prolactin influence mammary epithelial kinetics and differentiation and alter mouse mammary cancer risk. *Cancer Res. 53*: 1188–1194.

Engelman, R. W., Fukama, Y., Hamada, N., Good, R. A., & Day, N. K. (1991). Dietary restriction permits normal parturition and lactation but suppresses mouse tumor virus proviral transcription even after mammary involution. *Cancer Res. 51*: 5123–5128.

Everitt, A. V., Porter, B. D., & Wyndham, J. R. (1982). Effects of caloric intake and dietary composition on the development of proteinuria, age-associated renal disease and longevity in the male rat. *Gerontology 28*: 168–175.

Fernandes, G. & Good, R. A. (1984). Inhibition by restricted-calorie diet of lymphoproliferative disease and renal damage in MRL/lpr mice. *Proc. Natl. Acad. Sci. USA 81*: 6144–6148.

Fernandes, G., Chandrasekar, B., Troyer, D., Venkatraman, J. T., & Good, R. A. (1995). Dietary lipids and caloric restriction affect mammary tumor incidence and gene expression in mouse mammary tumor virus/v-Ha-ras transgenic mice. *Proc. Natl. Acad. Sci. USA 92*: 6494–6498.

Fernandes, G., Friend, P., Yunis, R. A., & Good, R. A. (1978a). Influence of dietary restriction on immunologic function and renal disease in (NZB×NZW)F$_1$ mice. *Proc. Natl. Acad. Sci. USA 75*: 1500–1504.

Fernandes, G., Rozek, M., & Troyer, D. (1986). Reduction of blood pressure and restoration of T-cell function in spontaneously hypertensive rats by food restriction and/or treadmill exercise. *J. Hypertens. 4*: (Suppl. 3) S469–S474.

Fernandes, G., Yunis, E. J., & Good, R. A. (1976). Influence of diet on survival of mice. *Proc. Natl. Acad. Sci. USA 73*: 1279–1283.

Fernandes, G., Yunis, E. J., & Good, R. A. (1989). Suppression of adenocarcinoma by immunologic consequences of caloric restriction. *Nature 263*: 504–507.

Fernandes, G., Yunis, E. J., Miranda, M., Smith, J., & Good, R. A. (1978b). Nutritional inhibition of genetically determined renal disease and autoimmunity with prolongation of life in *kdkd* mice. *Proc. Natl. Acad. Sci. USA 75*: 2888–2892.

Friend, P. S., Fernandes, G., Good, R. A., Michael, A. F., & Yunis, E. J. (1979). Dietary restrictions early and late – Effects on nephropathy of the NZB×NZW mouse. *Lab. Invest. 38*: 629–632.

Gong, X., Shang, F., Obin, M., Palmer, H., Scrofano, M. M., Jahgen-Hodge, J., Smith, D. E., & Taylor, A. (1997). Antioxidant enzyme activities in lens, liver, and kidney of calorie restricted Emory mice. *Mech. Ageing Dev. 99*: 181–192.

Gradin, K. & Persson, B. (1990). Blood pressure and sympathetic activity in spontaneously hypertensive rats during food restriction. *J. Neural Transm.* [Gen. Sect.] *79*: 183–191.

Grasl-Kraup, B., Bursch, W., Ruttkay-Nedecky, B., Wagner, A., Lauer, B., & Schulte-Hermann, R. (1994). Food restriction eliminates preneoplastic cells through apoptosis and antagonizes carcinogenesis in rat liver. *Proc. Natl. Acad. Sci. USA 91*: 9995–9999.

Gross, L. & Dreyfus, Y. (1984). Reduction in the incidence of radiation-induced tumors in rats after restriction of food intake. *Proc. Natl. Acad. Sci. USA 81*: 7596–7598.

Gross, L. & Dreyfus, Y. (1986). Inhibition of the development of radiation-induced leukemia in mice by reduction of food intake. *Proc. Natl. Acad. Sci. 83*: 7928–7931.

Gumprecht, L. A., Long, C. R., Soper, K. A., Smith, P. F., Haschek-Hock, W. M., & Keenan, K. P. (1993). The early effect of dietary restriction on the pathogenesis of chronic renal disease in Sprague-Dawley rats at 12 months. *Toxicol. Path. 21*: 528–537.

Hansen, B. C. & Bodkin, N. L. (1993). Primary prevention of diabetes mellitus by prevention of obesity in monkeys. *Diabetes 42*: 1809–1814.

Hansen, B. C., Bodkin, N. L., & Ortmeyer, H. K. (1999). Calorie restriction in non-human primates: Mechanisms of reduced morbidity and mortality. *Toxicol. Sci. 52*: (Suppl.) 56–60.

Hart, R., Dixit, R., Seng, J., Turturro, A., Leakey, J., Feuers, R., Duffy, P., Buffington, C., Cowan, G., Lewis, S., Pipkin, J., & Li, S. (1999). Adaptive role of caloric intake on degenerative disease processes. *Toxicol. Sci. 52*: *(2 Suppl)*: 3–12.

Heuson, J. C. & Legros, N. (1972). Influence of insulin deprivation on growth of the 7,12 dimethyl(α)anthracene-induced mammary carcinoma in rats subjected to alloxan diabetes and food restriction. *Cancer Res. 32*: 226–232.

Higami, Y., Yu, B. P., Shimokawa, I., Bertrand, H., Hubbard, G. B., & Masoro, E. J. (1995). Anti-tumor action of dietary restriction is lesion dependent in male Fischer 344 rats. *J. Gerontol.: Biol. Sci. 50A*: B72–B77.

Higami, Y., Yu, B. P., Shimokawa, I., Masoro, E. J., & Ikeda, T. (1994). Duration of dietary restriction: An important determinant for the incidence and age of onset of leukemia in male F344 rats. *J. Gerontol.: Biol. Sci. 49*: B239–B244.

Hursting, S. D., Perkins, F. N., & Phang, J. M. (1994). Caloric restriction delays spontaneous tumorigenesis in p53 knock-out transgenic mice. *Proc. Natl. Acad. Sci. USA 91*: 7036–7040.

Huseby, R. A., Ball, Z. B., & Visscher, M. B. (1945). Further observations on the influence of single calorie restriction on mammary cancer incidence and related phenomenon in C3H mice. *Cancer Res. 5*: 40–46.

Izui, S., Fernandes, G., Hara, I., McConahey, P. J., Jensen, F. C., Dixon, F. J., & Good, R. A. (1981). Low-calorie diet selectively reduces expression of retroviral envelope glycoprotein Gp70 in sera of NZB×NZWF1 hybrid mice. *J. Exp. Med. 154*: 1116–1124.

James, S. J. & Muskhelishvili, L. (1994). Rates of apoptosis and proliferation vary with caloric intake and may influence incidence of spontaneous hepatoma in C57BL/6 × C3HF1 mice. *Cancer Res. 54*: 5508–5510.

Jolly, C. A. & Fernandes, G. (1999). Diet modulates Th-1 and Th-2 cytokine production in the peripheral blood of lupus prone mice. *J. Clin. Immunol. 19*: 171–177.

Jolly, C. A., Muthukumar, A., Avula, C. P. R., Troyer, D., & Fernandes, G. (2001). Life span is prolonged in food-restricted autoimmune-prone (NZB×NZW)F(1) mice fed a diet enriched with (n-3) fatty acids. *J. Nutrition 131*: 2753–2760.

Jung, L. K. L., Palladino, M. A., Calvano, S., Mark, D. A., Good, R. A., & Fernandes, G. (1982). Effect of calorie restriction on the production and responsiveness to IL-2. *Clin. Immunol. Immunopath. 25*: 295–301.

Kealy, R. D., Lawler, D. F., Ballam. J. M., Lust, G., Smith, G. K., Biery, D. N., & Olsson, S. E. (1997). Five-year longitudinal study on limited food consumption and development of osteoarthritis in coxofemoral joints of dogs. *J. Am. Vet. Med. Assoc.* *210*: 222–225.

Keenan, K. P., Soper, K. A., Hertzog, P. R., Gumprecht, L. A., Smith, P. F., Mattson, B. A., Ballam, G. C., & Clark, R. L. (1995a). Diet, overfeeding, and moderate dietary restriction in control Sprague-Dawley rats: II. Effects on age-related proliferative and degenerative lesions. *Toxicol. Path. 23*: 287–302.

Keenan, K. P., Soper, K. A., Smith, P. F., Ballam, G. C., & Clark, R. L. (1995b). Diet, overfeeding, and moderate dietary restriction in control Sprague-Dawley rats: I. Effects on spontaneous neoplasms. *Toxicol. Path. 23*: 269–286.

Klurfeld, D. M., Weber, M. M., & Kritchevsky, D. (1987). Inhibition of chemically induced mammary and colon tumor promotion by caloric restriction in rats fed increased dietary fat. *Cancer Res. 47*: 2759–2762.

Klurfeld, D. M., Welch, C. B., Davis, M. J., & Kritchevsky, D. (1989a). Determination of degree of energy restriction necessary to reduce DMBA-induced mammary tumorigenesis in rats during the promotion phase. *J. Nutrition 49*: 286–291.

Klurfeld, D. M., Welch, C. B., Lloyd, L. M., & Kritchevsky, D. (1989b). Inhibition of DMBA-induced mammary tumorigenesis by caloric restriction in rats fed high-fat diets. *Internat. J. Cancer 43*: 922–925.

Koizumi, A., Tuskada, M., Hirano, S., Kamiyama, S., Masuda, H., & Suzuki, K. (1993). Energy restriction that inhibits cellular proliferation by torpor can decrease susceptibility to spontaneous and asbestos-induced lung tumors in A/J mice. *Lab. Invest. 68*: 728–739.

Koizumi, A., Wada, Y., Tsukada, M., Kamiyama, S., & Weindruch, R. (1990). Effects of energy restriction on mouse mammary tumor virus mRNA levels in mammary glands and uterus and on uterine endometrial hyperplasia and pituitary histology in CH3/SHN F1 mice. *J. Nutrition 120*: 1401–1411.

Koizumi, A., Wada, Y., Tsukada, M., Kayo, T., Naruse, M., Horiuchi, K., Mogi, T., Yoshioka, M., Sasaki, M., Miyamaura, Y., Abe, T., Ohtomo, K., & Walford, R. L. (1996). A tumor preventive effect of dietary restriction is antagonized by a high housing temperature through deprivation of torpor. *Mech. Ageing Dev. 92*: 67–82.

Koletsky, S. & Puterman, D. I. (1977). Reduction of atherosclerotic disease in genetically obese rats by low calorie diet. *Exp. Mol. Path. 20*: 415–424.

Kritchevsky, D. & Klurfeld, D. M. (1986). Influences of caloric intake on experimental carcinogenesis: A review. *Adv. Exp. Med. Biol. 206*: 55–68.

Kritchevsky, D., Weber, M. M., Buck, C. L., & Klurfeld, D. M. (1986). Calories, fat and cancer. *Lipids 21*: 272–274.

Kritchevsky, D., Weber, M. M., & Klurfeld, D. M. (1984). Dietary fat versus caloric content in initiation and promotion of 7, 12 dimethylbenz(a)anthracene-induced mammary tumorigenesis in rats. *Cancer Res. 44*: 3174–3177.

Kubo, C., Day, N. K., & Good, R. A. (1984). Influence of early or late dietary restriction on life span and immunological parameters of MRL/Mp-lpr/lpr mice. *Proc. Natl. Acad. Sci. USA 81*: 5831–5835.

Kubo, C., Gajjar, A., Johnson, B. C., & Good, R. A. (1992). The effects of dietary restriction on immune function and development of autoimmune disease in BXSB mice. *Proc. Natl. Acad. Sci. USA 89*: 3145–3149

Kumar, S. P., Roy, S. J., Tokumo, K., & Reddy, B. S. (1990). Effect of different levels of caloric restriction on azomethane-induced colon carcinogenesis in male F344 rats. *Cancer Res. 50*: 5761–5766.

Lane, M. A., Black, A., Ingram, D. K., & Roth, G. S. (1998). Calorie restriction in nonhuman primates: Implications for age-related disease risk. *J. Anti-Aging Med. 1*: 315–326.

Lane, M. A., Ingram, D. K., & Roth, G. S. (1999). Calorie restriction in nonhuman primates: Effects on diabetes and cardiovascular disease risk. *Toxicol. Sci. 52*: (Suppl.) 41–48.

Larsen, C. D. & Heston, W. E. (1945). Effects of cystine and caloric restriction on the incidence of spontaneous pulmonary tumors in strain A mice. *J. Natl. Cancer Inst. 6*: 31–40.

Lavik, B. S. & Bauman, C. A. (1943). Further studies on tumor promoting action of fat. *Cancer Res. 3*: 744–756.

Lewis, D. J. (1992). Non-neoplastic lesions in the cardiovascular system. In: U. Mohr, D. L. Dungworth, & C. C. Capen (Eds.), *Pathobiology of the Aging Rat*. Vol. I (pp. 301–309). Washington, DC: ILSI Press.

Lim, B. O., Jolly, C. A., Zaman, K., & Fernandes, G. (2000). Dietary (n-6) and (n-3) fatty acids and energy restriction modulate mesenteric lymph node lymphocyte function in autoimmune-prone (NZB×NZW)F1 mice. *J. Nutrition 130*: 1657–1664.

Lipman, R. D. (2002). Effect of calorie restriction on mortality kinetics in inbred strains of mice following 7,12-dimethylbenz[α]anthracene treatment. *J. Gerontol.: Biol. Sci. 57A*: B153–B157.

Lipman, R. D., Bronson, R. T., Chrisp, C., & Hazzard, D. (1996). Pathologic characterization of Brown Norway, Brown Norway×Fischer 344 and Fischer 344 × Brown Norway rats. *J. Gerontol.: Biol. Sci. 51A*: B54–B59.

Lipman, R. D., Dallal, G. E., & Bronson, R. T. (1999a). Lesion biomarkers of aging in B6C3F1 hybrid mice. *J. Gerontol.: Biol. Sci. 54A*: B466–B477.

Lipman, R. D., Dallal, G. E., & Bronson, R. T. (1999b). Effects of genotype and diet on age-related lesions in *ad libitum*-fed and calorie-restricted F344, BN, and BNF3F1 rats. *J. Gerontol.: Biol. Sci. 54A*: B478–B491.

Lipman, R. D., Smith, D. E., Blumberg, J. B., & Bronson, R. T. (1998). Effects of caloric restriction or augmentation in adult rats: Longevity and lesion biomarkers of aging. *Aging Clin. Exp. Res. 10*: 463–470.

Lloyd, T. (1984). Food restriction increases life span of hypertensive animals. *Life Sci. 34*: 401–407.

Luan, X., Zhao, B., Chandrasekar, B., & Fernandes, G. (1995). Caloric restriction modulates lymphocyte subset phenotype and increases apoptosis in MRL/lpr mice. *Immunol. Lett. 47*: 181–186.

Maeda, H., Gleiser, C. A., Masoro, E. J., Murata, I., McMahan, C. A., & Yu, B. P. (1985). Nutritional influences on aging of Fischer 344 rats: II. Pathology. *J. Gerontol. 40*: 671–688.

Manjgaladze, M., Chen, S., Frame, L. T., Seng, J. E., Duffy, P. H., Feuers, R. J., Hart, R. W., & Leakey, J. E. A. (1993). Effects of caloric restriction on rodent drug and carcinogen metabolizing enzymes: implications for mutagenesis and cancer, *Mutation Res. 295*: 201–222.

Masoro, E. J. (2001). Dietary restriction: An experimental approach to the study of the biology of aging. In: E. J. Masoro & S. N. Austad, (Ed.), *Handbook of the Biology of Aging*, 5th ed. (pp. 396–420). San Diego: Academic Press.

Masoro, E. J., Iwasaki, K., Gleiser, C. A., McMahan, C. A., Seo, E-J., & Yu, B. P. (1989). Dietary modulation of the progression of nephropathy in aging rats: an evaluation of the importance of protein. *Am. J. Clin. Nutr. 49*: 1217–1227.

Mattson, M. P., Duan, W., Lee, J., & Guo, Z. (2001). Suppression of brain aging and neurodegenerative disorders by dietary restriction and environmental enrichment: molecular mechanisms. *Mech. Ageing Dev. 122*: 757–778.

Morris, H. P. (1945). Some nutritional factors influencing the origin and development of cancer. *J. Natl. Cancer Inst. 6*: 1–11.

Mura, C. V., Roh, S., Smith, D., Palmer, V., Podhye, N., & Taylor, A. (1993). Cataract incidence and analysis of lens crystallins in the water-, urea-, and SDS-soluble fractions of Emory mice fed a diet restricted by 40% in calories. *Curr. Eye Res. 12*: 1081–1091.

Muskhelishvili, L., Turturro, A., Hart, R. W., & James, S. J. (1996). π-Class glutathione-S-transferase-positive hepatocytes in aging B6C3F1 mice undergo apoptosis induced by dietary restriction. *Am. J. Path. 149*: 1585–1591.

Muthukumar, A. R., Jolly, C. A., Zaman, K., & Fernandes, G. (2000). Calorie restriction decreases proinflammatory cytokines and polymeric Ig receptor expression in the submandibular glands of the autoimmune-prone (NZB×NZW)F$_1$ mice. *J. Clin. Immunol. 20*: 354–361.

Nolen, G. A. (1972). Effect of various restricted dietary regimens on the growth, health and longevity of albino rats. *J. Nutrition 102*: 1477–1494.

Notargiacomo, A. V. & Fries, E. D. (1981). Effect of weight-reducing diet on blood pressure of spontaneously hypertensive rats. *Proc. Soc. Exp. Med. Biol. 167*: 612–615.

Ogura, M., Ogura, H., Ikehara, S., Dao, M. L., & Good, R. A. (1989a). Decreases by chronic energy intake restriction of cellular proliferation in the intestinal epithelium and lymphoid organs in autoimmunity-prone mice. *Proc. Natl. Acad. Sci. USA 86*: 5918–5922.

Ogura, M., Ogura, H., Ikehara, S., & Good, R. A. (1989b). Influence of dietary restriction on the numbers and proportion of Lyn1$^+$ B lymphocytes in autoimmune prone mice. *Proc. Natl. Acad. Sci. USA 86*: 4225–4229.

Overton, J. M., Van Ness, J. M., & Casto, R. M. (1997). Food restriction reduces sympathetic support of blood pressure in spontaneously hypertensive rats. *J. Nutrition 127*: 655–660.

Pashko, L. I. & Schwartz, A. G. (1992). Reversal of food restriction-induced inhibition of mouse skin tumor promotion by adrenalectomy. *Carcinogenesis 13*: 1925–1928.

Pollard, M. & Luckert, P. M. (1985). Tumorigenesis effects of direct- and indirect-acting chemical carcinogenesis in rats on a restricted diet. *J. Natl. Cancer Inst. 74*: 1347–1349.

Pollard, M., Luckert, P. H., & Pan, G-Y. (1984). Inhibition of intestinal tumorigenesis in methylazomethanol-treated rats by dietary restriction. *Cancer Treat. Rept. 68*: 405–408.

Pollard, M., Luckert, P. H., & Snyder, D. (1989). Prevention of prostate and liver tumors in L-W rats by moderate dietary restriction. *Cancer 64*: 686–690.

Pugh, T. D., Oberley, T. D., & Weindruch R. (1999). Dietary intervention at middle age: caloric restriction but not dehydroepiandrosterone sulfate increases lifespan and life time cancer incidence in mice. *Cancer Res. 59*: 1642–1648.

Reddy, B. S., Wang, C-X., & Maruyama, H. (1987). Effect of restricted caloric intake on azomethane-induced colon tumor incidence in male F344 rats. *Cancer Res. 47*: 1226–1228.

Roe, F. J. C., Lee, P. N., Conybeare, G., Kelly, D., Mather, B., Prentice, D., & Tobin, G. (1995). The Biosure Study: Influence of composition of diet and food consumption on longevity, degenerative diseases, and neoplasia in Wistar rats studied for up to 30 months post weaning. *Food Chem. Toxicol. 33*: S1–S100.

Roebuck, B. D., Yager, Jr., J. D., & Longnecker, D. S. (1981a). Dietary modulation of azaserine-induced pancreatic carcinogenesis in the rat. *Cancer Res. 41*: 888–893.

Roebuck, B. D., Yager, Jr., J. D., Longnecker, D. S., & Wilpone, S. A. (1981b). Promotion by unsaturated fat of azaserine-induced pancreatic carcinogenesis in the rat. *Cancer Res. 41*: 3961–3966.

Rogers, A. E., Zeisel, S. H., & Groopman, J. (1993). Diet and carcinogenesis. *Carcinogenesis 14*: 2205–2217.

Ross, M. H. (1959). Protein, calories and life expectancy. Federation Proc. 18: 1190–1207.

Ross, M. H. & Bras, G. (1971). Lasting influence of early caloric restriction on the prevalence of neoplasms in the rat. *J. Natl. Cancer Inst. 47*: 1095–1113.

Ross, M. H., Lustbader, E. D., & Bras, G. (1983). Body weight, dietary practices, and tumor susceptibility in the rat. *J. Natl. Cancer Inst. 71*: 1041–1046.

Rous, F. (1914). The influence of diet on transplanted and spontaneous tumors. *J. Exp. Med. 20*: 433–451.

Rusch, H. P. (1944). Extrinsic factors that influence carcinogenesis. *Physiol. Rev. 24*: 177–204.

Safai-Kutti, S., Fernandes, G., Wang, Y., Safai, B., Good, R. A., & Day, N. K. (1980). Reduction of circulating immune complexes by calorie restriction in (NZB×NZW)F$_1$ mice. *Clin. Immunol. Immunopath. 15*: 293–300.

Sarkar, N. H., Fernandes, G., Telang, N. T., Kourides, J. A., & Good, R. A. (1982). Low-calorie diet prevents the development of mammary tumors in CH3 mice and reduces circulating prolactin level, murine mammary tumor virus expression, and proliferation of mammary alveolar cells. *Proc. Natl. Acad. Sci. USA 79*:7758–7762.

Saxton, J. A. Jr. & Kimball, G. C. (1941). Relation of nephrosis and other diseases of albino rats to age and to modifications of diet. *Arch. Path. 32*: 951–965.

Saxton, J. A. Jr., Boon, M. C., & Furth, J. (1944). Observations on the inhibition of development of spontaneous leukemia in mice by underfeeding. *Cancer Res. 4*: 401–409.

Seng, J. E., Gandy, J., Turturro, A., Lipman, R., Bronson, R. T., Parkinson, A., Johnson, W., Hart, R. W., & Leakey, J. E. A. (1996). Effects of caloric restriction on expression of testicular cytochrome P450 enzymes associated with the metabolic activation of carcinogens. *Arch. Biochem. Biophys. 335*: 42–52.

Sheldon, W. G., Bucci, T. J., Hart, R. W., & Turturro, A. (1995a). Age-related neoplasia in a lifetime study of *ad libitum*-fed and food restricted B6C3F1 mice, *Toxicol. Path. 23*: 458–476.

Sheldon, W. G., Bucci, T. J., & Turturro, A, (1996). Thoracic apophyseal osteoarthritis in feed-restricted and *ad libitum*-fed B6C3F1 mice. In: U. Mohr, D. Dungworth, C. Capen, W. Carlton, J. Sundberg, & J. Ward (Eds.), *Pathobiology of the Aging Mouse*, Vol. 1 (pp. 445–453). Washington, DC: ILSI Press.

Sheldon, W. G., Warbritton, A. R., Bucci, T. J., & Turturro, A. (1995b). Glaucoma in food restricted and ad *libitum fed*-DBA/2NNIA mice. *Lab. Anim. Sci. 45*: 508–518.

Shimokawa, I., Higami, Y., Hubbard, G. B., McMahan, C. A., Masoro, E. J., & Yu, B. P. (1993a). Diet and the suitability of the male Fischer 344 rat as a model for aging research. *J. Gerontol.: Biol. Sci. 48*: B27–B32.

Shimokawa, I., Yu, B. P., Higami, Y., Ikeda, T., & Masoro, E. J. (1993b). Dietary restriction retards onset but not progression of leukemia in male F344 rats. *J. Gerontol.: Biol. Sci. 48*: B68–B73.

Shimokawa, I., Yu, B. P., & Masoro, E. J. (1991). Influence of diet on fatal neoplastic disease in male Fischer 344 rats. *J. Gerontol.: Biol. Sci. 46*: B228–B232.

Silberberg, M. & Silberberg, R. (1955). Diet and life span. *Physiol. Rev. 35*: 347–362.

Sinhu, D. K., Gebhard, R. L., & Pazik, J. E. (1988). Inhibition of mammary carcinogenesis in rats by dietary restriction. *Cancer Lett. 40*: 133–141.

Snyder, D. L., Pollard, M., Worstmann, B. S., & Luckert, P. (1990). Life span, morphology, and pathology of diet-restricted germ-free and conventional Lobund-Wistar rats. *J. Gerontol.: Biol. Sci. 45*: B52–B58.

Stern, J. S., Gades, M. D., Wheeldon, C. M., & Borchers, A. T. (2001). Calorie restriction in obesity: Prevention of kidney disease in rodents. *J. Nutrition 131*: 913S–917S.

Stevens, H., Knollema, S., De Jong, G., Korf, J., & Luiten, P. (1998). Long-term food restriction, deprenyl, and nimodipine treatment on life expectancy and blood pressure of stroke-prone rats. *Neurobiol. Aging 19*: 273–276.

Susic, D., Mandal, A. K., Jovovic, D. J., Radujkovic, G., & Kentera, D. (1990). Streptozotocin-induced diabetes lowers blood pressure in spontaneously hypertensive rat. *Clin. Exp. Hypertens. A12*: 1021–1035.

Swoap, S. J., Bodell, P., & Baldwin, K. M. (1995). Interaction of hypertension and caloric restriction on cardiac mass and isomyosin expression. *Am. J. Physiol. 268*: R33–R39.

Sylvester, P. W., Aylsworth, C. F., & Meites, J. (1981). Relationship of hormones in inhibition of mammary tumor development by underfeeding during the "critical period" after carcinogen administration. *Cancer Res. 41*: 1384–1388.

Sylvester, P. W., Aylsworth, C. F., van Vogt, D. A., & Meites, J. (1982). Influence of underfeeding during "critical period" or thereafter on carcinogen-induced mammary tumors in rats. *Cancer Res. 42*: 4943–4947.

Tannenbaum, A. (1940). Initiation and growth of tumours. I. Effect of underfeeding. *Am. J. Cancer 38*: 335–350.

Tannenbaum, A. (1942). The genesis and growth of tumors. II. Effects of caloric restriction per se. *Cancer Res. 2*: 460–467.

Tannenbaum, A. (1944). The dependence of the genesis of induced skin tumors on the caloric intake during different stages of carcinogenesis. *Cancer Res. 4*: 673–677.

Tannenbaum, A. (1945). The dependency of tumor formation on the composition of the caloric-restricted diet as well as on the degree of restriction. *Cancer Res. 5*: 609–615.

Tannenbaum, A. & Silverstone, H. (1953a). Nutrition in relation to cancer. *Adv. Cancer Res. 1*: 451–501.

Tannenbaum, A. & Silverstone, H. (1953b). Effects of limited food intake on survival of mice bearing spontaneous mammary carcinoma and on the incidence of lung metastases. *Cancer Res. 13*: 532–536.

Taylor, A. Lipman, R., Jahngen-Hodge, J., Palmer, V., Smith, D., Padhye, N. Dallal, G. E., Cyr, D. E., Laxman, E., Shepherd, D., Morrow, F., Salomon, R., Perrone, G., Asmundsen, G., Meydani, M., Blumberg, J., Mune, M., Harrison, D. E., Archer, J. R., & Shigenaga, M. (1995). Dietary caloric restriction in the Emory mouse: Effects on lifespan, eye lens cataract, prevalence and progression, levels of ascorbate, glutathione, glucose, glycohemaglobin, tail collagen break time, DNA and RNA oxidation, skin integrity, fecundity, and cancer. *Mech. Ageing Dev. 79*: 33–57.

Taylor, A., Zuliani, A. M., Hopkins, R. E., Dallal, G. E., Treglia, P., Kuck J. E. & Kuck, K. (1989). Moderate calorie restriction delays cataract formation in the Emory mouse. *FASEB J. 3*: 1741–1746.

Thurman, J. D., Bucci, T. J., Hart, R. W., & Turturro, A. (1994). Survival, body weight, and spontaneous neoplasms in *ad libitum*-fed and food restricted Fischer 344 rats. *Toxicol. Path. 22*: 1–14.

Thurman, J. D., Greenman, D. L., Kodel, R. L., & Turturro, A. (1997). Oral squamous cell carcinoma in *ad libitum*-fed and food-restricted Brown-Norway rats. *Toxicol. Path. 25*: 217–224.

Thurman, J. D., Moeller, jr., R. B., & Turturro. A. (1995). Proliferative lesions of the testis in *ad libitum*-fed and food-restricted Fischer 344 and FBNF$_1$ rats. *Lab. Animal Sci. 45*: 635–640.

Troyer, D. A., Chandrasekar, B., Barnes, J. L., & Fernandes, G. (1997). Calorie restriction decreases platelet-derived growth factor (PDGF)-A and thrombin receptor mRNA expression in autoimmune lupus nephritis. *Clin. Exp. Immunol. 108*: 58–62.

Troyer, D. A., Chandrasekar, B., Thinnes, T., Stone, A., Loskutoff, D. A., & Fernandes, G. (1995). Effects of energy intake on type I plasminogen activator inhibitor levels in glomeruli of lupus-prone B/W mice. *Am. J. Pathol. 146*: 111–120.

Tucker, M. J. (1979). Effect of long-term food restriction on tumours in rodents. *Internat. J. Cancer 23*: 802–807.

Tucker, S. M., Mason, R. L., & Beauchene, R. E. (1976). Influence of diet and food restriction on kidney function in aging male rats. *J. Gerontol. 31*: 264–270.

Turturro, A., Blank, K., Murasko, D., & Hart, R. (1994). Mechanism of caloric restriction affecting aging and disease. *Ann. N Y Acad Sci. 719*: 159–170.

Verdery, R. B. & Walford, R. L. (1998). Changes in plasma lipids and lipoproteins in humans during a 2-year period of dietary restriction in Biosphere 2. *Arch. Intern. Med. 158*: 900–906.

Verdery, R. B., Ingram, D. K., Roth, G. S., & Lane, M. A. (1997). Caloric restriction increases HDL$_2$ levels in rhesus monkeys *(Macaca mulatta)*. *Am. J. Physiol. 273*: E714-E719.

Vischer, M. B., Ball, Z. B., Barnes, R. H., & Silverstein, I. (1942). The influence of caloric restriction upon the incidence of spontaneous mammary carcinoma in mice. *Surgery 11*: 48–55.

Volk, M. J., Pugh, T. D., Kim, M. J., Frith, C. H., Daynes, R. A., Ershler, W. B., & Weindruch, R. (1994). Dietary restriction from middle age attenuates age-associated lymphoma development and interleukin dysregulation in C57BL/6 mice. *Cancer Res. 54*: 3054–3061.

Walford, R. L., Harris, S. B., & Gunion, M. W. (1992). The calorically restricted low-fat nutrient-dense diet in Biosphere 2 significantly lowers blood glucose, total leukocyte count, cholesterol, and blood pressure in humans. *Proc. Natl. Acad. Sci. USA 89*: 11533–11537.

Walford, R. L., Mock, D., Verdery, R., & MacCallum (2002). Caloric restriction in Biosphere 2: Alterations in physiologic, hematologic, hormonal, and biochemical parameters in humans restricted for a 2-year period. *J. Gerontol.: Biol. Sci. 57A*: B211–B224.

Warner, H. R., Fernandes, G., & Wang, E. (1995). A unifying hypothesis to explain the retardation of aging and tumorigenesis by caloric restriction. *J. Gerontol.: Biol. Sci. 50A*: B107–B109.

Weindruch, R. (1989). Dietary restriction, tumors, and aging in rodents. *J. Gerontol. 44*: (Special Issue) 67–71.

Weindruch, R. & Walford, R. L. (1982). Dietary restriction in mice beginning at 1 year of age: Effect on life-span and spontaneous cancer incidence. *Science 215*: 1415–1418.

White, F. R. (1961). The relationship between underfeeding and tumor formation, transplantation, and growth in rats and mice. *Cancer Res. 21*: 281–290.

White, F. R., White, J., Mider, G. B., Kelly, M. G., Heston, W. E., & David, P. W. (1944). Effect of caloric restriction on mammary tumor formation in strain C3H mice and on the response to painting with methylcholanthrene. *J. Natl. Cancer Inst. 5*: 43–48.

Wolf, N. S., Li, Y., Pendergrass, W., Schmeider, C., & Turturro, A. (2000). Normal mouse and rat strains as models for age-related cataract and the effect of caloric restriction on its development. *Exp. Eye Res. 70*: 683–692.

Wright, G. L., McMurty, J. P., & Wexler, B. C. (1981). Food restriction reduces blood pressure of the spontaneously hypertensive rat. *Life Sci. 28*: 1253–1259.

Yu, B. P., Masoro, E. J., & McMahan, C. A. (1985). Nutritional influences on aging of Fischer 344 rats. I. Physical, metabolic, and longevity characteristics. *J. Gerontol. 40*: 657–670.

Yu, B. P., Masoro, E. J., Murata, I., Bertrand, H. A., & Lynd, F. T. (1982). Life span study of SPF Fischer 344 male rats fed *ad libitum* or restricted diets: Longevity, growth, lean body mass and disease. *J. Gerontol. 37*: 130–141.

CHAPTER 6

Mechanisms of the anti-aging action of caloric restriction

Contents

The mechanism underlying life-extension and other anti-aging effects of CR have long been a major interest of biological gerontologists. Although many mechanisms have been proposed, empirical support for most of them is weak or lacking, and for some, the experimental evidence tends to invalidate the concept.

In their initial report, McCay et al. (1935) proposed that CR increases longevity of rats by retarding growth. As discussed in Chapter 2, this hypothesis must be eliminated because the anti-aging actions of CR, including life-extension, were found to occur even when CR is initiated in adult life of rats and mice.

Berg and Simms (1960) suggested that CR retards aging by reducing body fat content. In Chapter 4, evidence was presented that makes this hypothesis unlikely. However, the recent availability of technology for noninvasive measurements of individual fat depots will enable further evaluation of the hypothesis. This is certainly warranted in light of the data showing that increased levels of abdominal visceral fat heighten the risk of age-associated diseases in humans (Masoro, 2001).

Sacher (1977) proposed that CR retards aging by reducing the metabolic rate. As discussed in Chapter 4, there is evidence that CR need not decrease the intensity of metabolism per unit of "metabolic mass" to bring about its anti-aging and life-prolonging actions. Nevertheless, this

RESEARCH PROFILES IN AGING
VOLUME 1 ISSN 1567-7184

hypothesis continues to have strong supporters. Indeed, the ultimate resolution of this issue will require testing the hypothesis by measuring the metabolic rate of specific organs, tissues and even cell types under usual living conditions. Technology is now being developed that may permit such measurements to be conducted in a noninvasive fashion in the not too distant future.

Lowering the ambient temperature, and thus the body temperature of poikilothermic species, has been found to increase their life span (Finch, 1990). The findings from studies showing that CR lowers the body temperature of mice, rats, and rhesus monkeys were reviewed in Chapter 4, and the possibility that the reduction in body temperature plays an important part in the anti-aging actions of CR was considered. Based on the currently available information, it was concluded that the reduction in body temperature does not appear to play a major role in CR's anti-aging actions.

It has been suggested that CR may retard aging by increasing physical activity (McCarter, 2000). However, as discussed in Chapter 4, CR does not always increase the level of physical activity. In a critical review of the literature, McCarter (2000) concludes that the anti-aging actions of CR are not the result of an increase in physical activity.

There have been many hypotheses based on the effects of CR on particular cellular or biochemical processes, such as the enhancement of apoptosis which increases the elimination of damaged cells (Zhang & Herman, 2002) and the increased turnover of body protein, which steps up elimination of damaged proteins (Tavernakis & Driscoll, 2002). As of this writing, experiments aimed at testing these cellular and biochemical hypotheses have not yielded supportive evidence or have not been done.

However, when our current knowledge of the physiological actions and gene expression effects of CR is combined with the findings on genetic manipulations that extend life span of vertebrate and invertebrate species, four hypotheses emerge; and these hold promise of yielding a mechanistic understanding of the anti-aging action of CR. I refer to these hypotheses by the following names: Oxidative Damage Attenuation; Alteration of the Glucose–Insulin System; Alteration of the Growth Hormone-IGF-1 Axis; and Hormesis. The remainder of this chapter will focus on an in-depth discussion of these four hypotheses.

Oxidative damage attenuation

Reactive oxygen molecules, such as hydroxyl radicals, superoxide radicals, and hydrogen peroxide, cause oxidative damage in important molecules (lipids, proteins, and nucleic acids) in living cells. Reactive oxygen molecules are generated by intrinsic living processes, such as the processes of energy metabolism, and by environmental factors, such as cigarette smoke. The damaging action of these molecules is referred to as oxidative stress, and it has been proposed that oxidative stress underlies much of senescent deterioration and that CR's anti-aging action stems from its ability to attenuate oxidative stress. Thus, the name Oxidative Damage Attenuation Hypothesis is appropriate. Although several investigators have proposed this hypothesis using various other names, a paper by Sohal and Weindruch (1996) is particularly useful because it provides a clear and succinct presentation of the hypothesis.

There, is, indeed, an age-associated buildup of oxidatively damaged molecules in the tissues of rodents, and CR has been found to retard this accumulation (Yu, 1996). Matsuo et al. (1993) reported that CR suppresses the age-associated increase in the rate of exhalation of ethane and pentane in rats, indicating that CR attenuates the increase with age in the rate of lipid peroxidation. In fact, CR slows the age-associated accumulation of peroxidized lipids in the liver, brain, heart, kidneys and other organs of rats (Baek et al., 1999; Cook & Yu, 1998; Kim et al., 1995; Laganiere & Yu, 1987; Pieri et al., 1992; Rao et al., 1990; Rikans et al., 1991) and mice (Chipalkatti et al., 1983; Davis et al., 1993; Immre & Juhaz, 1987; Koizumi et al., 1987). Figure 6-1, taken from Cook and Yu (1998), illustrates the effects of age and CR on lipid peroxidation in the kidney, liver, and brain of male F344 rats. CR also decreases the accumulation of lipofuscin in both rats and mice (Enesco & Kruk, 1981; Rao, et al., 1990).

CR also decreases the age-associated increase in oxidatively damaged proteins in rats (Aksenova et al., 1998; Youngman et al., 1992) and in mice (Dubey et al., 1996; Sohal et al., 1994b). It does so by attenuating the increase in the carbonylation of proteins (Figure 6-2) and the loss of protein sulfhydryl groups (Figure 6-3). CR of 6 weeks duration, initiated in *ad libitum*-fed old mice, results in a reduction of protein oxidative damage, reducing the carbonyl content and increasing the sulfhydryl content, in brain, although the effect is not equivalent to that of long-term CR; however, CR initiated in late-life does not affect the sulfhydryl content of the heart (Forster et al., 2000). CR prevents the age-associated increase in tyrosyl radical-induced oxidative dityrosine cross-linking of cardiac and skeletal muscle proteins in mice (Leeuwenburgh et al., 1997).

CR protects mice and rats from the age-associated accumulation of oxidatively damaged DNA (Chen & Snyder, 1992; Chung et al., 1992;

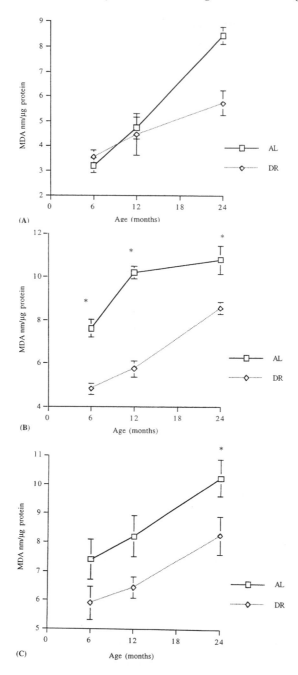

Figure 6-1. Age, CR and lipid peroxidation in the tissues of male F344 rats. (A) kidneys; (B) liver; (C) brain. AL refers to *ad libitum*-fed rats and DR to rats on long-term CR. (From Cook and Yu, 1998.)

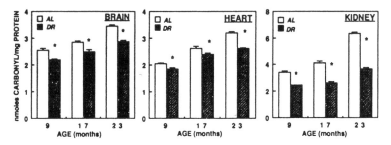

Figure 6-2. Effects of age and CR on the carbonyl content of proteins of the brain, heart, and kidney of male C57BL/6 mice. AL denotes *ad libitum*-fed mice and DR denotes mice on CR from 4 months of age. (From Sohal et al., 1994b.)

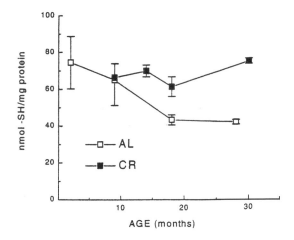

Figure 6-3. Effects of age and CR on the sulfhydryl content of mitochondrial proteins in the skeletal muscle of male C57BL/6 mice. AL denotes *ad libitum*-fed mice. (From Lass et al., 1998.)

Sohal et al., 1994a; Youngman, 1993). Data are presented graphically in Figure 6-4 on the extent of DNA oxidation in the tissues of male 15-month-old C57BL/6 mice that were either fed *ad libitum* or on CR starting at 4 months of age (Sohal et al., 1994a). The effects of age and CR on the hepatic DNA oxidation in mice is shown in Figure 6-5 from Sohal et al. (1994a) and the effect of CR on hepatic nuclear and mitochondrial DNA oxidation in 24-month-old male F344 rats is reported in Figure 6-6 from Chung et al. (1992). CR also decreases oxidative damage to the mitochondrial DNA in the hearts of rats (Gredilla et al., 2001). Somewhat surprisingly, CR increases the urinary

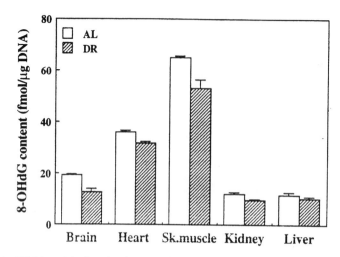

Figure 6-4. DNA oxidation in tissues of 15-month-old male C57BL/6 mice fed *ad libitum* or on CR starting at 4 months of age. DR denotes CR. (From Sohal et al., 1994a.)

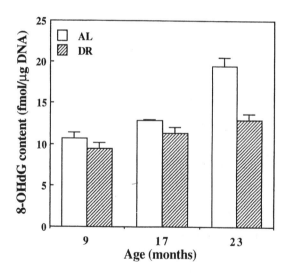

Figure 6-5. Effects of age and CR on the oxidation of hepatic DNA in male C57BL/6 mice. DR denotes CR. (From Sohal et al., 1994a.)

excretion of oxidation products of DNA and RNA in Emory mice (Taylor et al., 1995). Recently, Hamilton et al. (2001) pointed out that much of the research on DNA oxidation may have yielded artifactual information due to the use of analytic methods that cause DNA

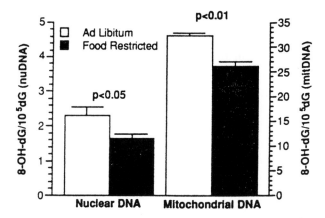

Figure 6-6. Nuclear and mitochondrial DNA oxidation in the liver of 24-month-old F344 rats fed *ad libitum* or on CR starting at 6 weeks of age. Food restricted denotes CR. (From Chung et al., 1992.)

oxidation. They developed methodology that circumvents this problem and confirmed that CR reduces age-associated accumulation of oxidatively damaged DNA in all tissues of male B6C23F1 mice and most tissues of male F344 rats; it also decreases the accumulation of oxidatively damaged mitochondrial DNA in the liver of mice and rats.

CR has been found to prevent the age-associated accumulation of oxidative damage in mouse skeletal muscle mitochondria (Lass et al., 1998). Mitochondrial proton leak-dependent oxygen consumption in skeletal muscle is decreased in very old rats maintained on CR starting at 10 months of age (Lal et al., 2001); this finding is important because increases in mitochondrial proton leak may be both a cause and a consequence of oxidative stress.

Studies on the effects of age and CR on oxidative damage have mostly utilized mice or rats as the animal model. However, oxidative damage has also been found to occur with increasing age in the skeletal muscle of rhesus monkeys, and CR of 10 years duration has been found to decrease the extent of this damage (Zainal et al., 2000).

This attenuation of the increase in oxidatively damaged molecules means that CR either decreases the rate of generation of reactive oxygen molecules, or it increases the effectiveness of protective and repair processes, or it does both. The production of superoxide, hydroxyl radicals and hydrogen peroxide is less in hepatic mitochondria and microsomes from rats on CR than in those from *ad libitum*-fed animals (Lee & Yu, 1990). Similar findings were found with rat brain mitochondria (Choi & Yu, 1994). CR also decreases the generation of reactive oxygen species by the postmitochondrial fraction of brain of 24-month-old rats (Baek et al., 1999). Long-term CR decreases reactive oxygen species production by rat heart mitochondria; this decrease occurs at Complex I of the electron transport chain and is not due to a diminution of mitochondrial oxygen consumption but rather to a lower degree of reduction of the Complex I generator, which decreases its percentage of free radical leak (Gredilla et al., 2001). CR has also been found to decrease the rate of generation of superoxide radicals and hydrogen peroxide by mitochondria from the brain, kidney, and heart of mice (Sohal & Dubey, 1994; Sohal et al., 1994b). Interestingly, a 50% reduction in food intake for one week in young *ad libitum*-fed male Sprague-Dawley rats markedly decreases the mRNA level of Uncoupling Protein-3 in skeletal muscle (Boss et al., 1998); extension of this kind of research in an aging study is sorely needed. CR decreases cyclooxygenase activity and the generation of cyclooxygenase-derived reactive oxygen species in the brain and kidney of rats (Baek, et al., 2001; Chung et al., 1999). The above findings on decreased generation of reactive oxygen species are all from *in vitro* studies and, as pointed out by Feuers et al. (1993), little is known about the effects of CR on *in vivo* free radical production. It is, indeed, risky to draw conclusions about *in vivo* processes based solely on *in vitro* findings.

The influence of CR on antioxidant defenses has proven to be extremely complex. Many studies, particularly the early reports, indicated that CR in mice and rats increases the activities and expression of antioxidant enzymes and/or retards the age-associated decrease in the activities of these enzymes (Chen et al., 1996; Heydari & Richardson, 1992; Kim et al., 1995; Koizumi et al., 1987; Laganiere & Yu, 1989; Pieri et al., 1992; Rao et al., 1990; Richardson, 1991; Semsei et al., 1989). However, a number of other studies have failed to consistently demonstrate that CR enhances the activities of these enzymes (Gong et al., 1997; Mote et al., 1991; Mura et al., 1996; Rojas et al., 1993; Sohal et al., 1994b). Indeed, in one study, CR was found to attenuate the age-associated increase in rat skeletal muscle antioxidant enzyme activities (Luhtala et al., 1994). And Dhahbi et al. (1998) found that in the female mouse, aging leads to an increase in translational efficiency of hepatic catalase mRNA, while CR obviates this effect. Xia et al. (1995) used male F344 rats to study the effects of age and CR (started soon after weaning) on the activity of antioxidant enzymes; cerebral cortex, heart, intestinal mucosa, kidney and liver were assessed at 6, 16, and 26 months of age for catalase, glutathione peroxidase, and Cu, Zn superoxide dismutase activities. CR did not affect the intestinal mucosa activities of these enzymes at any of the ages studied. In the heart, CR increased: catalase activity at all ages studied; glutathione peroxidase activity at 16 and 26 but not 6 months of age; and Cu, Zn superoxide dismutase activity at 6 and 16 but not 26 months of age. In the kidney, CR increased: catalase activity at 16 and 26 but not 6 months of age; glutathione peroxidase activity at 26 months of age but not at the younger ages; and Cu, Zn superoxide dismutase activity at all ages studied. In the liver, CR increased: catalase and glutathione peroxidase activities at 26 months of age but not at younger ages; and Cu, Zn superoxide dismutase activity at 16 and 26 but not 6 months of age. In the cerebral cortex, CR increased catalase and Cu, Zn superoxide dismutase activities at 26 months of age, but not at the other ages studied; and it did not affect glutathione peroxidase activity at any age studied. However, Baek et al. (1999) found that catalase activity decreases with age in the cerebellum of rats, and CR prevents this decrease. The conclusion to be drawn from the studies carried out to date is that many factors influence the effects of CR on the activities of the antioxidant enzymes; these include age, gender, and tissue specificities, and this complexity may explain what appear to be inconsistencies in the findings reported in the literature.

Moreover, antioxidant enzymes exhibit a complex circadian pattern of activity. Oriaku et al. (1997) studied this circadian pattern in the livers of young female F344 rats, which had been fed *ad libitum* or on a CR regimen for 6 weeks; they concluded that CR enhances the rat's enzymatic ability to remove free radicals generated as a consequence of normal oxidative metabolism. This study also suggests one of the reasons for lack agreement on CR's effects on antioxidant enzyme activity; that is, the failure of investigators to consider the circadian pattern of activity and the effect of CR on that pattern.

Furthermore, a complication in the study of catalase activity stems from the fact that it is inactivated by reactive oxygen species, including its substrate, hydrogen peroxide. Also, Feuers et al. (1997) found that when CR is started at about 3 months of age, it reduces the accumulation of inactive catalase in the liver of 9- to 12-month-old male (BN × F344) F_1 rats. This finding indicates that CR may increase the capacity to remove reactive oxygen species; but it also suggests that assessment of this enzymatic activity may be affected by inactivation by its substrate during the assay. So here is another possible factor for the discordant findings in the literature on the influence of CR on calalase activity.

CR is known to bolster some nonenzymatic antioxidant defenses. Cellular reduced glutathione (GSH) levels decrease with increasing age, and CR has been found to increase, into advanced ages, the levels in rat and mouse livers, particularly the mitochondrial glutathione level (Armeni et al., 1998; Chen et al., 1996; Laganiere & Yu, 1989; Taylor et al., 1995). Figures 6-7 and 6-8 from the study of Armeni et al.

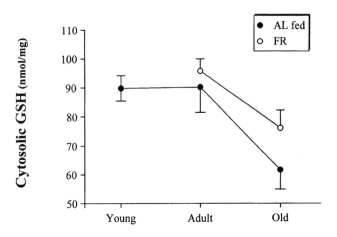

Figure 6-7. Effects of age and CR on the cytosolic concentration of GSH in the liver of female Wistar rats. Young denotes 6-month-old, adult denotes 11-month-old, and old denotes 24-month-old. AL fed denotes *ad libitum*-fed and FR denotes CR. (From Armeni et al., 1998.)

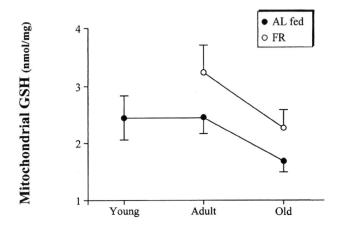

Figure 6-8. Effects of age and CR on the mitochondrial concentration of GSH in the liver of female Wistar rats. Young denotes 6-month-old, adult denoted 11-month-old, and old denotes 24-month-old. AL fed denotes *ad libitum*-fed and FR denotes CR. (From Armeni et al., 1998.)

(1998) show the effects of age and CR on the hepatic cytosolic and mitochondrial GSH concentrations, respectively. This action may provide a major defense against oxidative damage. Also, Cook and Yu (1998) have shown that CR blunts the age-associated increase in the iron content of the kidney, brain and liver of rats (Figure 6-9); this effect may well attenuate lipid peroxidation. CR increases the hepatic level of uric acid in male F344 rats, possibly providing protection against oxidative damage to this organ (Chung & Yu, 2000). In Emory mice, CR causes a marked decrease in the level of ascorbate in plasma, liver, and kidney (Taylor et al., 1995); in regard to this, it should be noted that ascorbate can function as a prooxidant as well as an antioxidant.

Indeed, there is direct evidence that CR can protect against the damaging actions of exposure to reactive oxygen molecules. Long-term CR has been found to protect mouse lens epithelial cells from damage caused by *in vitro* exposure to hydrogen peroxide (Li et al., 1998). Also, CR protects mitochondrial gene transcription from the damaging effect of peroxyl radicals (Kristal & Yu, 1998).

As discussed in Chapter 3, CR enhances the repair of oxidatively damaged DNA molecules, while Chapter 4 covers the promotion of protein turnover by CR. Clearly, rodents on CR have an increased ability to replace damaged proteins with newly synthesized undamaged molecules.

Based on all these studies, there is little question that CR protects rats and mice from oxidative stress. Does this mean that this protective action is the major or sole mechanism underlying the anti-aging and life-extending actions of CR? In order to answer this question, it is necessary to know the role of oxidative stress in the aging of these organisms. Is the accumulation of oxidative damage the cause of aging, or is it merely one of the many phenotypic results of fundamental aging processes? There is some evidence suggesting the former. In fruit flies selected for postponed senescence, there is an increased resistance to oxidative stress (Arking et al., 1991); however, it is not known whether the delay in aging in these long-lived genotypes is specifically due to a resistance to oxidative stress, or to a resistance to stressors in general, or to a mechanism or mechanisms not related to stress. Also, the life span of mice is markedly increased by the homozygous mutation of p66[shc] gene, and these mice show no obvious phenotypic defect (Migliaccio et al., 1999). The authors suggest that the increase in longevity is due to an increased cellular and organismic resistance to oxidative stress because the adaptor protein coded for by the p66[shc] gene is part of a signal transduction pathway activated by increased levels of intracellular reactive oxygen species.

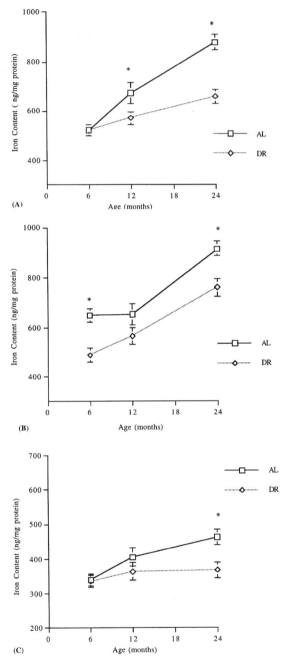

Figure 6-9. The effects of age and CR on the iron content in the kidney (panel A), the liver (panel B), and the brain (panel C) of male F344 rats. AL refers to *ad libitum*-fed rats and DR to rats on long-term CR. (From Cook and Yu, 1998.)

However, it is important to remember that such a mutation may have many consequences in addition to the response to oxidative stress. Therefore, although there is reason to feel that oxidative stress may play a major role in aging and that its attenuation may be a mechanism by which CR retards aging, direct evidence in support of these concepts is not yet in hand.

Alteration of the glucose-insulin system

Sustained elevation of plasma glucose and insulin levels is damaging to mammalian organisms (Reaven, 1989), and this damage has some similarities to what occurs during senescence (LevRan, 1998; Parr, 1996). Thus, it is possible that normal levels of plasma glucose and insulin, acting slowly over the course of a lifetime, can cause much of the deterioration that is characteristic of the human senescent phenotype as well as that of other mammalian species. CR alters many characteristics of rodent and nonhuman primate carbohydrate metabolism, including the plasma levels of glucose and insulin. Based on these effects, the Alteration of the Glucose-Insulin System Hypothesis was proposed as a mechanistic explanation of the anti-aging and life-prolonging action of CR (Masoro, 1996).

Early studies on the effects of CR on the glucose-insulin system are difficult to interpret, because they did not provide lifelong information on the levels of plasma glucose and insulin and the use of glucose fuel under usual daily living conditions. Reaven and Reaven (1981) measured plasma glucose and insulin levels once a day (between 2 and 4 PM) in male Sprague-Dawley rats on either a low or high energy intake. They found that between 3 and 12 months of age, plasma glucose levels were not affected by energy intake, but plasma insulin levels were lower in the rats on the lower energy intake. In several studies, the effects of CR on the fasting levels of plasma glucose and/or insulin were determined in mice and rats, and it was found that CR reduces the level of both (Harris et al., 1994; Koizumi et al., 1989; Masoro et al., 1983; Reaven et al., 1983; Ruggeri et al., 1989). Belage et al. (1990) reported that CR decreases the postprandial level of plasma insulin in male Sprague-Dawley rats. Reaven et al. (1983) studied glucose utilization in anesthetized 12- to 13-month-old male Sprague-Dawley rats fasted for 4 hours with a method involving the continuous infusion of epinephrine, propranolol, glucose, and insulin; they concluded that the rats on a lower caloric intake are less insulin-resistant than those on a higher caloric intake. However, this finding was not confirmed by Kalant et al. (1988) who studied the effect on glucose utilization of CR started at 4 weeks of age in male F344 rats, using the same method as Reaven et al. Glucose utilization was measured in 4-, 12-, 18- and 24-month-old rats after an overnight fast; CR resulted in a decrease in the ability of insulin to stimulate glucose utilization at all ages studied. Ivy et al. (1991) studied glucose metabolism in a hind limb perfusion preparation obtained from anesthetized Long-Evans male rats that had been fasted overnight; those preparations from CR rats exhibited increased glucose utilization compared to the *ad libitum*-fed rats.

Masoro et al. (1992) carried out a longitudinal study on male F344 rats to determine plasma glucose and insulin concentrations and glucose utilization over a lifetime under usual daily living conditions. The circadian pattern of plasma glucose concentration was measured (Figure 6-10). Over most of the 24 hours, rats on CR had lower plasma glucose levels than those fed *ad libitum*; only during the 2 to 3 hours following their daily feeding did the plasma glucose level in the rats on CR approach that of the rats fed *ad libitum*. Over their lifetime, the CR rats sustained an approximately 15% reduction in mean 24-hour level of plasma glucose (Table 6-1). In this study, plasma insulin levels were measured during the daily periods of maximum and minimum plasma glucose levels (Figure 6-11); at both times, the level of plasma insulin was markedly lower in CR rats than in *ad libitum*-fed rats. Using male F344 rats maintained in the same facility and on the same dietary regimens, McCarter and Palmer (1992) carried out a lifetime longitudinal study of the oxygen consumption and respiratory quotient in two rat groups, one on the CR regimen and the other fed *ad libitum*. Daily oxygen consumption per kg of "metabolic mass" or lean body mass was the same over most of the life span for both dietary groups as was the respiratory quotient of 0.89, a value indicating that the rats were using a daily mixture of fuels identical to the mixture in their diet, which is not surprising for rats in a near steady state in regard to body mass and composition. In addition, another set of 8–10-month-old CR and *ad libitum*-fed rats were studied during the 2-hour period in which both dietary groups reached their daily maximal plasma glucose concentration; during these 2 hours, the following were measured: plasma glucose and insulin levels, oxygen consumption per kg "metabolic mass," and respiratory quotient (Masoro et al., 1992). There was no difference between the two groups in oxygen consumption, respiratory quotient, and plasma glucose concentration, but the plasma insulin concentration in the CR rats was about 50% that of the *ad libitum*-fed rats. The conclusion drawn from the above study is that the CR rats were able to daily utilize a similar amount of glucose fuel per kg "metabolic mass" as the *ad libitum*-fed rats, while maintaining a lower mean 24-hour plasma glucose

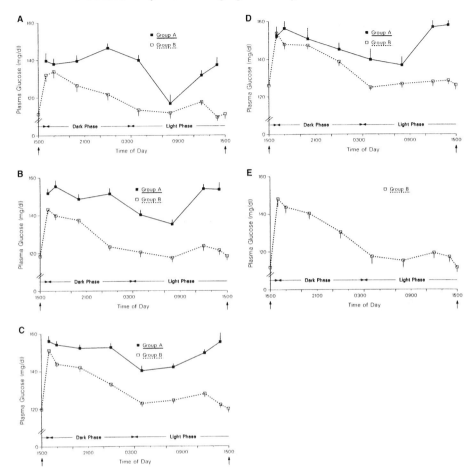

Figure 6-10. Circadian pattern of plasma glucose concentration in *ad libitum*-fed (Group A) and CR (Group B) male F344 rats. Graphs: A, 3–7 months of age; B, 9–13 months of age; C, 15–19 months of age; D, 21–25 months of age; and E, 27–31 months of age. Vertical arrows indicate the time at which the Group B rats received the daily allotment of food. (From *J. Gerontol.: Biol. Sci. 47*: B202–B208, 1992; Copyright © The Gerontological Society of America. Reproduced by permission of the publisher.)

concentration and a markedly lower plasma insulin concentration. Thus, it is clear that CR increases either glucose effectiveness, or insulin sensitivity, or both.

Some research has been done on the mechanisms underlying the effects of CR on carbohydrate metabolism in rodents. *In vitro* studies show that CR enhances insulin stimulation of glucose transport in rat skeletal

Table 6-1
Effect of CR on 24-hour mean plasma glucose concentration in male F344 rats

Age range, months	AL mg/dl	CR mg/dl
3–7	136 (2)	119 (2)
9–13	147 (2)	126 (2)
15–19	149 (2)	131 (2)
21–25	148 (3)	135 (3)
27–31		127 (3)

Note: AL denotes *ad libitum*-fed; numbers in parentheses denote SE; no measurements were made on AL rats in the age range of 27–31 months. (Data from Masoro et al., 1992.)

muscle (Cartee et al., 1994; Davidson, 1978; Dean & Cartee, 1996). Cartee and Dean (1994) found that this enhancement began as soon as 5 days after the initiation of CR (a 25% reduction in caloric intake) in the rat epitrochlearis muscle but not in the flexor digitorum brevis muscle. *In vitro* studies also show that CR increases glucose effectiveness in rat skeletal muscle (Mc Carter et al., 1994). The glucose transporter GLUT-4 accounts for most of the insulin-stimulated glucose transport in skeletal muscle; although CR does not alter the level of GLUT-4 in rat skeletal muscle, it increases the fraction of the GLUT-4 located in the plasma membrane of the muscle cell during insulin stimulation, i.e., the component of GLUT-4 that functions in glucose transport (Dean et al., 1998a). In the same muscle study, CR did not alter insulin's stimulation of insulin receptor substrate (IRS)-1-associated phophatidylinositol-3-kinase (PI3K) activity, nor the abundance of IRS-2 and p85 subunit of PI3K; but it did lower the abundance of IRS-1. Davidson et al. (2002) also reported that CR of 20-days duration increases glucose transport by rat skeletal muscle, and this is not associated with changes in IRS-1-, IRS-2-, or phophotyrosine-phophatidylinositol-3-kinase activity. CR also enhances insulin stimulation of glucose transport in mouse skeletal muscle, and IRS-1 is not involved in this action (Gazdag et al., 1999). Balage et al. (1990) reported that CR does not enhance insulin binding by rat skeletal muscle; this does not agree with the study by Wang et al. (1997) who found that CR increases insulin binding to its receptor in rat skeletal muscle, though not in heart.

In vitro studies have shown that CR enhances insulin's ability to promote glucose metabolism in adipocytes (Craig et al., 1987). It also enhances insulin binding by isolated adipocytes of rats (Olefsky & Reaven, 1975).

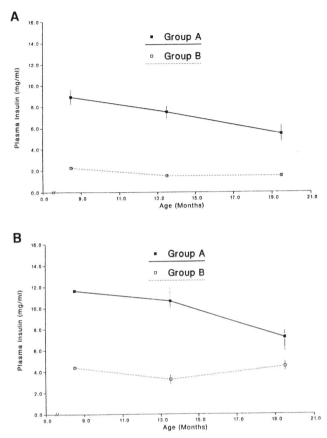

Figure 6-11. Longitudinal study of plasma insulin concentration in Group A (*ad libitum*-fed) and CR (Group B) male F344 rats. Graph A reports plasma insulin concentration at the time of the daily minimum plasma glucose concentration and Graph B reports plasma insulin concentration at the time of the daily maximum plasma glucose concentration. (From *J. Gerontol.: Biol. Sci. 47*: B202–B208, 1992; Copyright © The Gerontological Soeicty of America. Reproduced by permission of the publisher.)

CR influences hepatic insulin action and carbohydrate metabolism in rodents. Barzilai and Gabriely (2001) found that CR increases the ability of insulin to suppress hepatic glucose production in both young and old rats. In contrast, Escriva et al. (1992) reported that CR decreases the ability of insulin to suppress hepatic glucose production in young rats. In a study by Kalant et al. (1988), CR did not influence insulin binding or insulin degradation by rat liver, while in the study of Balage et al. (1990), CR enhanced insulin binding by rat liver. Insulin receptor mRNA levels are increased in the liver of mice on a CR regimen (Spindler et al., 1991). CR decreases the enzymatic capacity for glycolysis in mouse liver and increases it for gluconeogensis (Dhahbi et al., 1999).

Escriva et al. (1992) used the tritium-labeled 2-deoxyglucose-procedure to carry out an *in vivo* study on the effect of CR on glucose metabolism. They studied anesthetized 70-day-old Wistar rats under basal and hyperinsulinemic conditions, and found that CR increased insulin-mediated glucose uptake by various skeletal muscle and adipose tissue sites. Wetter et al. (1999) did a similar study using 11–12-month old F344×BN F_1 rats, and it is important to note that they carried out the study in as close to usual living conditions as possible. They found that although CR maintained low levels of plasma insulin, it did not affect glucose uptake by cerebellum, lung, kidney, soleus muscle or diaphragm, and it enhanced glucose uptake by brown adipose tissue and white adipose tissue as well as the epitroclearis, plantaris, and gastrocnemius muscles.

Several studies indicate that the markedly lower plasma insulin levels of rats on CR are due largely to decreased secretion of insulin by the β-cells of the pancreatic islets (Bergamini et al., 1991; Dean et al., 1998b; Rao, 1995; Reaven et al., 1983). However, there is also evidence that CR may increase insulin clearance in rodents (Feuers et al., 1995).

Long-term CR was found to decrease fasting plasma glucose and insulin levels in two studies on rhesus monkey (Kemnitz et al., 1994; Lane et al., 1995). In another rhesus monkey study, CR decreased fasting plasma insulin levels but did not affect fasting plasma glucose concentration (Bodkin et al., 1995). Findings in all three studies led the authors to conclude that CR increases insulin sensitivity in rhesus monkeys. In the study of Kemnitz et al. (1994), insulin sensitivity was determined with the Modified Minimal Model method; Lane et al. (1995) determined insulin sensitivity by analyzing the glucose tolerance test; and Bodkin et al. (1995) used the euglycemic, hyperinsulinemc clamp technique. Moreover, fasting plasma insulin levels and the insulin response in an intravenous glucose tolerance test were significantly decreased in old rhesus monkeys (greater than 18 years of age) on a CR regimen for only 12 months (Lane et al., 1999). In all the rhesus monkey studies, the measurement of plasma glucose and insulin and the determination of insulin sensitivity were done on fasting animals under light anesthesia. Thus, these studies provide no information on effects of CR on glucose and insulin levels and glucose utilization in rhesus monkeys under normal living conditions. Cefalu et al. (1997) found that after one year on a CR regimen, adult cynomolgus monkeys exhibit an increase in insulin sensitivity, as determined by the

Modified Minimal Model method. Fasting blood sugar is decreased in humans after 6 months on a low-calorie diet (Walford et al., 1992).

CR does not affect the abundance of the GLUT-4 glucose transporter, nor the abundance of two components of the insulin signaling pathway, phosphatidylinositol-3 kinase p85 subunit and insulin receptor substrate-1, in skeletal muscle of rhesus monkeys (Gazdag et al., 2000). CR does affect, in unexpected ways, rhesus monkey skeletal muscle glycogen metabolism (Ortmeyer et al., 1994, 1998). It increases basal glycogen synthase activity but prevents insulin activation of this enzyme, and it increases glycogen phosphorylase activity and the level of intracellular glucose 6-phosphate in response to insulin.

In regard to the anti-aging action of CR, investigators initially focused on the sustained reduction of plasma glucose levels during the life of an organism. There are two reasons for the initial focus. First, chronic hyperglycemia has been linked to microvascular disorders, macrovascular disease, basement membrane thickening, impaired cellular immunity, and cell cycle abnormalities (Rossetti et al., 1990). Thus, it is reasonable to believe that the normoglycemic level may slowly do the same, and that lowering it would be protective. Second, Cerami (1985) proposed the glycation theory of aging. It is based on the occurrence of nonenzymatic glycation of protein and nucleic acid molecules and subsequent reactions to yield advanced glycation end products (AGE). It is further postulated that this results in damage to these macromolecules as well as cellular damage caused by the AGE alteration of the macromolecules and thereby aging of the organism. Moreover, it is well established that elevated levels of plasma glucose will promote glycation and thus the generation of AGE (Cerami, 1985; Monnier, 1990). In addition, glycation is associated with oxidative reactions (Dunn et al., 1989) and, thus, relates to the oxidative stress hypothesis (Kristal & Yu, 1992).

Several studies have shown that CR influences glycation and AGE accumulation in rodents. Rats and mice on a CR regimen have lower glycation levels of hemoglobin and plasma proteins than control rats (Cefalu et al., 1995; Masoro, et al., 1989; Sell, 1997); the findings of Sell are summarized in Table 6-2. CR decreases plasma fructosamine levels in male and female F344 rats (Van Liew et al., 1993). It slows the age-associated increase in the glycation of rat and mouse skin collagen (Sell, 1997) and reduces the accumulation of glycoxidation products, N^{ε}-(carboxymethyl)lysine and pentosidine, in rat skin and tendon collagen (Cefalu et al., 1995; Sell et al., 1996). Sell and Monnier (1997) found that CR retarded the age-associated increases in pentosidine in the tendon and

Table 6-2
Effect of CR on the level of glycated hemoglobin in male C57BL/6 mice

Age, months	% Glycated hemoglobin	
	AL	CR
12	3.40(0.4)	2.64(0.2)
18	3.17(0.3)	2.58(0.3)
24	2.18(0.4)	2.33(0.1)

Note: AL denotes *ad libitum*-fed; numbers in parentheses denote SE. (Data from Sell, 1997.)

ear auricle of DBA/2 mice, though the effects of CR were less evident in C57BL/6 mice. On the other hand, Reiser (1994) reported that CR decreases the age-associated increase of AGE in skin and tendon collagen of C57BL/6 mice. As discussed in Chapter 3, there have been conflicting findings on the effects of age and CR on protein glycation and related reactions, which has somewhat tempered enthusiasm for the concept that attenuating glycation plays a major role in the anti-aging action of CR.

The focus has recently shifted to the reduction in plasma insulin levels as a mediator of CR's anti-aging action. Elevated levels of plasma insulin have long been implicated as a contributor to age-associated human cardiovascular disease, such as atherosclerosis, coronary heart disease, and hypertension (Ducimetiere et al., 1980; Ferrari & Weidman, 1990; Stout, 1990). However, the main reason for the new shift in focus is the research on genetic manipulations that significantly extend the life span of the nematode *C. elegans*; loss-of-function mutations in genes coding for components of the cellular insulin-like signaling pathway markedly increase the longevity of these worms (Lane, 2000; Vanfleteren & Braeckman, 1999). Indeed, it has been suggested that an evolutionary link connects the loss-of-function mutations in the nematode and the life-extending action of CR (Kimura et al., 1997; Lin et al., 1997). A loss-of-function mutation in an insulin-like signaling pathway has also been found to extend the life span of *Drosophila melanogaster* (Clancy et al., 2001; Tatar et al., 2001). In addition, Guarente's group has used a medium with a low glucose concentration as a way of increasing the longevity of yeast; presumably this procedure mimics caloric restriction in rodents (Lin et al., 2000). They found that the increase in longevity does not occur in the yeast mutant for *SIR-2*, the gene that encodes for the silencing protein Sir2p. Subsequently, Tissenbaum and Guarente (2001) reported that increased dosage of the *SIR-2* gene extends the life span of *C. elegans*, and they suggest that silencing by Sir2 may couple nutrient availability to the level of signaling through the insulin-like signaling pathway of the nematode.

"Negative" was my first reaction to the suggestion that these findings in nematodes, fruit flies, and yeast may relate to the anti-aging action of CR in rodents since the insulin-signaling system in skeletal muscle cells of rodents is not decreased, but rather is enhanced by CR. However, the work of Wolkow et al. (2000) caused me to reconsider this view. They restored the functioning of the insulin-signaling pathway in individual tissues of a loss-of-function mutated strain of *C. elegans*; and they found that only when they eliminated loss-of-function in the nervous system (but not in muscle and intestine) do the mutant nematodes lose the mutation-induced increase in longevity. Thus, if CR enhances the insulin-signaling system pathway in skeletal muscle but not in the nervous system of rodents, then the low plasma insulin levels in CR rodents could have an effect on insulin signaling in the nervous system of rats and mice that is similar to that of the loss-of-function mutation in nervous system insulin-like signaling of nematodes.

In summary, there is, indeed, evidence that CR decreases glycation and related reactions. Such findings lend support to the Alteration of the Glucose-Insulin System hypothesis as a mechanism in the anti-aging action of CR. However, whether glycation and related reactions play an important role in senescence has yet to be established. In addition, even if they do play an important role, it is an open question as to whether the extent to which CR influences these reactions is sufficient to have a marked effect on aging. The evidence that insulin signaling plays an important role in senescence of nematodes and fruit flies is compelling. In fact, many invertebrate biologists enthusiastically embrace the concept that there is a universal component to aging in animal organisms, and that this component may be insulin signaling. However, as yet the data in support of this concept are far from robust. Nevertheless, both the glycemia component and the insulinemia component of the Alteration of the Glucose-Insulin System hypothesis remain viable and certainly merit further study.

Alteration of the growth hormone-IGF-1 axis

CR influences the functioning of the growth hormone-IGF-1 axis, and this action may play an important role in its life-prolonging and anti-aging actions. The research on both Snell and Ames dwarf mice is the major reason for advancing this view. Such dwarf mice have life spans that are significantly longer than their normal sized littermates

(Brown-Borg et al., 1996; Flurkey et al., 2002; Miller, 1999). Although the Snell and Ames dwarf mice have mutations in different genes, they exhibit similar phenotypic characteristics (Bartke et al., 2001a); in addition to being dwarfs, both exhibit underdeveloped anterior pituitary glands that are devoid of cells producing growth hormone, prolactin, and thyroid stimulating hormone. Mice of the long-lived $(C3H/HeJ \times DW/J)F_1$ background with a defect in growth hormone production, due to a loss-of-function mutation at the *Pit-1* locus, exhibit a greater than 40% increase in mean and maximal longevity (Flurkey et al., 2001). Based on the work of Coschigano et al. (2000), it appears likely that the increase in longevity in the Snell and Ames dwarf mice results from the lack of growth hormone and the concomitant decrease in peripheral levels of IGF-1. These investigators studied mice with a targeted disruption of the growth hormone receptor/growth hormone binding protein gene, and they found that such animals had a significant extension of life span along with a decrease in body size and low levels of plasma IGF-1.

I refer to the concept that the anti-aging actions of CR are due to its effect on the growth hormone-IGF-1 axis as the Alteration of the Growth Hormone-IGF-1 Axis hypothesis. Further support for proposing this hypothesis is the fact that nematodes and fruit flies do not have separate insulin and IGF-1 signaling pathways; rather, both signaling pathways in mammals are homologues of the single signaling pathway in these invertebrate species. Thus, the genetic studies on nematodes and fruit flies, which suggest a possible role of insulin signaling in senescence, are equally applicable to the mammalian IGF-1 signaling pathway.

Indeed, CR markedly alters the growth-hormone-IGF-1 axis in rats and mice. CR for 10 weeks suppresses the pulsatile secretion of growth hormone in rats over a wide range of ages (Armario et al., 1987; Quigley et al., 1990). A 40% or 60% reduction in food intake for 30 days markedly decreases the serum concentration of growth hormone in young male Sprague-Dawley rats, and the 60% restriction also lowers the serum concentration of growth hormone binding protein (Oster et al., 1995). Growth hormone secretion decreases markedly with increasing age in BN rats, and although CR suppresses growth hormone secretion in the young rats, old rats on a long-term CR regimen have a secretory activity similar to that of young *ad libitum*-fed rats (Sonntag et al., 1995). Also, CR of 3 months duration, initiated in BN rats at 26 months of age, restores their growth hormone secretory activity to that of young *ad libitum*-fed rats (Sonntag et al., 1999). Long-term CR increases the number of somatotropes in the anterior pituitary of old rats (Shimokawa et al., 1996); it modulates somatotrope cell turnover, thereby preserving the cell population for growth hormone secretion in rats of advanced ages (Shimokawa et al., 1997b). It also slightly enhances the pituitary sensitivity to growth hormone-releasing hormone (Shimokawa et al., 2000). Long-term but not short-term CR protects against the age-related loss of pituitary high affinity growth hormone-releasing hormone binding sites (Girard et al., 1998). Growth hormone secreting cells from old F344 rats, studied *in vitro*, showed an increased sensitivity to somatostatin-28, and long-term CR attenuated this age-associated increase in sensitivity (Shimokawa et al., 1997a). However, CR does not affect plasma growth hormone levels in rhesus monkeys (Cocchi et al., 1995).

In mice, a progressive impairment in the growth hormone signal transduction occurs with increasing age, and CR delays this functional impairment (Xu & Sonntag, 1996a). In male B6D2 mice, growth hormone-induced activation of hepatic stat-3 decreases with increasing age; and this contributes to the age-associated impairment in hepatic growth hormone signal transduction and IGF-1 gene expression. CR attenuates this decrease resulting in an enhancement in the hepatic response to growth hormone and an increase in IGF-1 secretion (Xu & Sonntag, 1996b).

In young adult Sprague-Dawley rats, a 60% reduction in food intake for 30 days lowers the hepatic IGF-1 mRNA level and markedly decreases plasma IGF-1 concentration; a 40% reduction in food intake for the same length of time has a similar though more modest effect on plasma IGF-1 concentration but does not affect the hepatic IGF-1 mRNA level (Oster et al., 1995). CR also markedly decreases plasma IGF-1 levels in young BN rats (D'Costa et al., 1993), but with increasing age, there is a fall in the plasma IGF-1 levels in the *ad libitum*-fed rats and not in rats on the CR regimen (Figure 6-12). Indeed, at advanced ages, plasma IGF-1 concentration is about the same in both dietary groups (Breese et al., 1991). Similar findings on plasma IGF-1 levels were observed with mice (Sonntag et al., 1992). Although Tomita et al. (2001) found that CR decreases plasma IGF-1 levels in adult F344 rats, an age-associated decrease was not seen in *ad libitum*-fed rats in this rat strain. CR was found to decrease plasma IGF-1 levels in young but not in old rhesus monkeys (Cocchi et al., 1995). In aging rats, CR increases the density of IGF Type I receptors in liver, heart and skeletal muscle by some 1.5–2.5 fold (D'Costa et al., 1993). Tomita et al. (2001) found that CR increases the level of IGF Type I receptor mRNA in skeletal muscle of F344 rats at 6 and 16 months of age and the receptor protein at

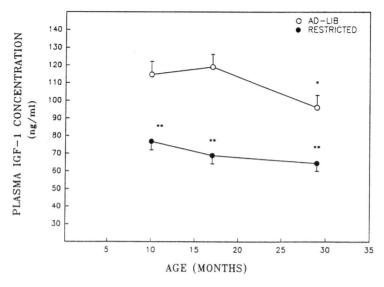

Figure 6-12. Effects of age and CR on plasma IGF-1 concentration in male BN rats. AD-LIB denotes *ad libitum*-fed rats and RESTRICTED denotes CR rats. (From D'Costa et al., 1993.)

6 months of age; it also maintains the skeletal muscle IGF-1-mRNA and insulin receptor substrate-1-mRNA at levels similar to those in *ad libitum*-fed F344 rats. Tomita et al. concluded that CR may enable the rat to maintain local tissue levels of IGF-1 and the components of its signaling pathway at levels similar to those of *ad libitum*-fed rats while lowering plasma levels. Moreover, there is evidence that autocrine/paracrine IGF-1 function may be as, or even more, important than its hormone function (Yakar et al., 1999).

It is extremely difficult to interpret these many findings as either supporting or negating the Alteration of the Growth Hormone-IGF-1 Axis hypothesis. Bartke et al. (2001b) point out that the relationships among growth hormone, growth, maturation, body size, and aging are complex and in need of further study. A guide for further study is provided by a recent conjecture of Bartke and Lane (2001). The following is a paraphrase of their idea: During young life, a high level of growth hormone and IGF-1 stimulates growth and metabolism, but this may also act to decrease longevity (e.g., the larger the breed of dog, the shorter its life span). However, at advanced ages, these hormones may retard aging by preventing adiposity, the loss of lean body mass, and the deterioration of the vascular system. Kalu et al. (1998) found that the administration of growth hormone, starting at 17 months of age in Balb/c mice and 18 months of age in F344 rats (late middle age for both species), had no effect on longevity. This finding does not entirely support the speculation of Bartke and Lane since the late-life elevation of growth hormone did not increase longevity, but it is significant that neither did it decrease longevity. The latter is reassuring in that the administration of exogenous growth hormone, for instance, to p^{53}-deficient mice on a CR regimen restores their plasma level of IGF-1 to that of *ad libitum*-fed mice, though it decreases the ability of CR to retard p-cresidine-induced bladder cancer in these mice (Dunn et al., 1997). Such findings raised concerns that increasing plasma growth hormone and IGF-1 levels at advanced ages might lead to a decrease in longevity by fostering neoplasia and other age-associated disease processes; the study of Kalu et al. indicates this need not be a concern. Although CR rodents and Ames dwarf mice have many characteristics in common, they also have characteristics that differ. Because of these differences, Mattison et al. (2000) concluded that the Ames dwarf mice cannot be considered to be CR mimetics. While this conclusion seems sound, it does not rule out the possibility that one or more of the characteristics shared by the two animal models underlie the increase in longevity that both exhibit. However, Bartke et al. (2001c) reported that CR further extends the life span of the Ames dwarf mice and based on this finding, they concluded that the slowing of aging by mutations in the growth hormone-IGF-1 axis involves a different mechanism than that of CR. Clancy et al. (2002) presented evidence that the findings reported by Bartke et al. do not necessarily support such a conclusion. Although there is not yet strong evidence in its support, the Alteration of the Growth Hormone-IGF-1 Axis hypothesis clearly warrants further exploration.

Hormesis

Hormesis refers to the phenomenon whereby a usually detrimental chemical substance or environmental agent changes its role to provide beneficial effects when administered at low concentrations or intensities (Furst, 1987). Although many different definitions of hormesis have been proposed, the following is useful in the context of gerontology: *Hormesis is the beneficial action resulting from the response of an organism to a low-intensity stressor*. It is important to note that in the aging context, *beneficial action* refers to increasing longevity, slowing senescent deterioration, retarding age-associated diseases and coping with high intensity stressors; however, it must also be noted that other aspects of the organism's biology may be affected adversely (e.g., a decrease in fecundity) or not affected at all. In 1998, I proposed that CR's life-extending and anti-aging actions in rodents are the result of hormesis (Masoro, 1998). Turturro et al. (1998) also concluded that the effect of CR on rodent survival is an example of hormesis.

The first issue that must be addressed is whether CR, as utilized in experimental biological gerontology, is a chronic, low-intensity stressor. Strong evidence that CR is, indeed, a chronic low-intensity stressor comes from the fact that rats on a lifelong CR regimen maintain a daily modest elevation in the afternoon peak plasma concentration of free corticosterone (Sabatino et al., 1991); the life span circadian pattern of plasma free corticosterone in male F344 rats is shown in Figure 6-13. The study of Amario et al. (1987) showed that CR for 34 days increases the daily level of plasma total corticosterone in approximately 3-month-old male Sprague-Dawley rats. Stewart et al. (1988) confirmed the finding that CR increases plasma total corticosterone in young rats (5-month-old male F344 rats), but found that it does not increase plasma total corticosterone levels in 24-month-old rats. Sabatino et al. (1991) also found that the increase in plasma total corticosterone in young rats on CR disappears with increasing age, but that the daily peak in plasma free corticosterone levels remains elevated in CR rats throughout life. This is important because the plasma free corticosterone level is believed to be the component of functional importance (Mendel, 1989). Young mice on CR also show an elevation in the daily peak concentration of plasma corticosterone; however, a life span study of this species has yet to be done (Klebanov et al., 1995). Han et al. (1995) found that the CR-induced increase in peak plasma concentration of corticosterone in rats does not relate to an increase in plasma adrenocorticotropic hormone (ACTH) concentration; and they speculated that the underlying mechanism is an increase in the responsiveness of the adrenal cortex to ACTH. In a subsequent study, such was found to be the case for young rats (Han et al., 1998). However, when Han et al. (2001) recently extended their research to include old rats, they found that while CR through middle age increases the sensitivity of the adrenal cortex to ACTH, the increased sensitivity wanes at older ages, even though the peak afternoon concentration of plasma free corticosterone continues to remain elevated. Although the underlying basis of this action of CR in old rats remains to be identified, it must be noted that a lifetime characteristic of CR rats is the maintenance of high levels of plasma free corticosterone. The fact that two

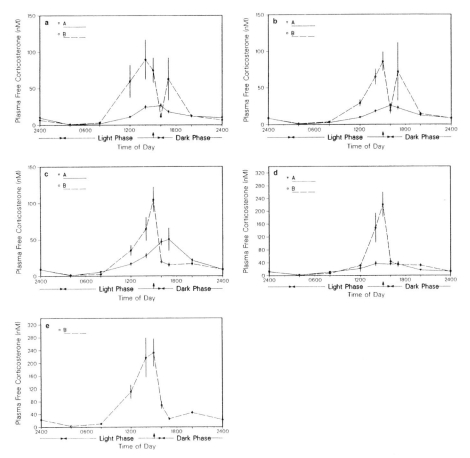

Figure 6-13. Circadian pattern of plasma free corticosterone concentration in *ad libitum*-fed (A) and CR (B) rats. Age-ranges: a, 3–7 months; b, 9–13 months; c, 15–19 months; d, 21–25 months; e, 27–31 months. Vertical arrows denote the time at which the CR rats received the daily food allotment. (From *J. Gerontol.: Biol. Sci. 46*: B171–B179, 1991; Copyright © The Gerontological Society of America. Reproduced by permission of the publisher.)

or more mechanisms are used to sustain the elevated levels of plasma free corticosterone throughout the life of CR rats is a strong indication of its functional importance.

The next issue is whether rodents on a CR regimen exhibit hormesis when coping with damaging agents (i.e., intense stressors). There is evidence that CR protects rodents of all ages from damage caused by acute, intense stressors. As is evident from Figure 6-14, CR attenuates the acute loss in body weight of rats following surgical stress

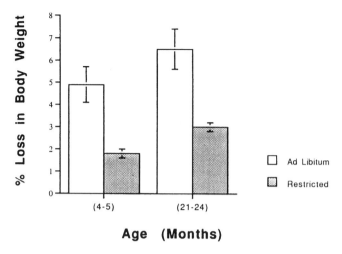

Figure 6-14. Effect of CR on the percent loss of body weight during the 48 hours following surgical implantation of a jugular canula. Numbers in parentheses refer to age of male F344 rats in months. Restricted denotes rats on CR. From Masoro (1998).

Table 6-3
Effect of CR on the survival of 20-month-old male F344 rats after heat stress

Dietary regime	Survival	
	Number: alive/dead	%
AL	15/77	16
CR	27/9	75

Note: Data from Heydari et al. (1993).

(Masoro, 1998). In addition, it reduces the inflammatory reaction seen in mice following injection of carageenan in the footpad (Klebanov et al., 1995). Heydari et al. (1993) found that rats on CR have an increased ability to survive a sudden marked increase in environmental temperature (Table 6.3). Also, CR protects rodents from damage caused by the action of toxic drugs (Duffy et al., 1995); e.g., following the administration of ganciclovir sodium, CR markedly reduces the mortality rate of female B6C3F$_1$ mice (Berg et al., 1994). Further, Keenan et al. (1997) tested four pharmaceutical drug candidates on Sprague-Dawley rats fed either *ad libitum* or a CR regimen, and found that CR rats needed higher doses to achieve classical maximum tolerated doses.

Since rodents on CR exhibit hormesis in coping with acute, intense stressors, the question arises regarding the relevance of this action to CR's effects on aging and longevity. Many biological gerontologists, including this author, believe that a major component of senescence in all animal species stems from the failure to repair damage or fully prevent long-term action of damaging agents (low-intensity stressors) generated by intrinsic living processes as well as by extrinsic agents. If this concept is correct, then the same hormetic processes that enhance coping with acute, intense stressors should also retard aging processes and increase longevity. Indeed, the effect of hormesis on longevity was demonstrated many years ago by Maynard Smith (1958) who reported a large increase in the life span of female fruit flies when subjected to transient heat shock. Since then, there have been many other examples of longevity hormesis resulting from exposure to low levels of a toxic agent (Johnson et al., 1996). Moreover, genetic manipulations that yield genotypes with increased longevity often result as well in an increase in the ability of the organism to cope with acute, intense stressors (Lithgow & Walker, 2002; Martin et al., 1996). As discussed above, such has been shown for *C. elegans* (Johnson et al., 2000), *D. melanogaster* (Lin et al., 1998), and yeast (Fabrizio et al., 2001). Kapahi et al. (1999) carried out an *in vitro* study on the ability of fibroblasts of eight mammalian species to cope with stressors in relation to the life span of the species. They found a positive correlation between the ability of the fibroblast to cope with a variety of stressors and species life span. These many diverse findings are supportive of the hypothesis that hormesis may be the basis for the life-prolonging action of CR and possibly its anti-aging action. A caveat in regard to the latter: Although longevity and senescence are related, they are not in lock step.

The final question: What are the organismic, cellular, and molecular processes that are utilized by hormesis to extend life and retard aging? The effect of CR on plasma free corticosterone levels is likely to be involved. It is well known that the hypothalamic-hypophyseal-adrenal cortical glucocorticoid system plays an important role in the ability of mammals to cope with damage (Munck et al., 1984). Thus, it seems reasonable to suggest that the elevated maximal daily levels of plasma free corticosterone enable the rodent to better cope with the daily damaging actions of intrinsic processes and extrinsic agents that ultimately result in senescence. In support of this speculation it has been demonstrated that CR's ability to protect mice against chemically induced tumors is lost if the animals have been adrenalectomized (Pashko & Schwartz, 1992). Indeed, Schwartz and Pashko (1994) have proposed that elevated levels of adrenal steroids underlie the ability of CR to inhibit both carcinogen-induced cancers as well as those occurring spontaneously with increasing age. In addition, Birt et al. (1999) have presented evidence that CR inhibits skin carcinogenesis in mice through the mediation of elevated levels of glucocorticoids. Also, Leakey et al. (1994) have postulated that the increased levels of corticosterone contribute to the anti-aging actions of CR.

In addition to this organismic hormonal mechanism, it is likely that hormesis involves cellular processes that protect against damage and those that repair damage. Families of stress-response genes, which are common to a broad spectrum of species, function to protect cells from the damaging actions of stressors (Papaconstaninou et al., 1996). Indeed, CR has been found to influence at least some stress-response genes in that it enhances the expression of heat shock proteins in response to stressors (Aly et al., 1994, Heydari et al., 1993, 1995; Moore et al., 1998; Pipkin et al., 1994). Also, it must be regarded as significant that a transgenic strain of *D. melanogaster* that has an increased ability to induce hsp 70 shows an increase in longevity (Tatar et al., 1997).

Although many of CR's actions clearly fit the concept of hormesis, Neafsey (1990) concluded that the increased longevity of rats resulting from CR is not due to hormesis. Her conclusion is based on differences between the effects of CR and those of longevity hormesis (induced by methylene chloride and gamma radiation) on the age-associated increase in age-specific mortality. However, it must be noted that in drawing this conclusion, Neafsey used a different and narrower definition of hormesis than the one proposed in this chapter. It should also be noted that the hormesis hypothesis is in accord with current views on the evolution of the anti-aging actions of CR, as discussed in Chapter 7. What is now needed is a clear delineation of the processes that link CR as a low-intensity stressor to CR's role as a protector against intense stressors, and thus its role in the retardation of aging processes. Until this is forthcoming, the hormesis hypothesis must remain just one of several promising hypotheses.

References

Aksenova, M. V., Aksenov, M. Y., Carney, J. M., & Rutterfield, D. A. (1998). Protein oxidation and enzyme activity decline in old Brown Norway rats are reduced by dietary restriction. *Mech. Ageing Dev.* 100: 157–168.

Aly, K. B., Pipkin, J. L., Hinson, W. G., Feuers, R. J., Duffy, P. R., Lyn-Cook, L., & Hart, R. (1994). Chronic caloric restriction induces stress proteins in the hypothalamus of rats. *Mech. Ageing Dev.* 76: 11–23.

Arking, R., Buck, S., Berrios, A., Dwyer, S., & Baker, G. T. (1991). Elevated paraquat resistance can be used as a bioassay for longevity in a genetically based long-lived strain of Drosophila. *Dev. Genet. 12*: 362–370.

Armario, A., Montero, J. L., & Jolin, T. (1990). Chronic food restriction and the circadian rhythms of pituitary-adrenal hormones, growth hormone, and thyroid-stimulating hormone. *Ann. Nutr. Metab. 31*: 81–87.

Armeni, T., Pieri, C., Marra, M., Saccucci, F., & Principato, G. (1998). Studies on the life prolonging effect of food restriction: Glutathione levels and glyoxylase enzymes in rat liver. *Mech. Ageing Dev. 101*: 101–110.

Baek, B. S., Kim, J. W., Lee, J. I., Kwon, H. J., Kim, N. D., Kang, H. S., Yoo, M. A., Yu, B. P., & Chung, Y. (2001). Age-related increase of brain cyclooxygenase activity and dietary modulation of oxidative status. *J. Gerontol.: Biol. Sci. 56A*: B426–B431.

Baek, B. S., Kwon, H. J., Lee, K. L., Yoo, M. A., Kim, K. W., Ikeno, Y. & Yu, B. P. (1999). Regional differences of ROS generation, lipid peroxidation, and antioxidant enzyme activity in rat brain and their dietary modulation. *Arch. Pharm. Res. 22*: 361–366.

Bartke, A. & Lane, M. (2001). Endocrine and neuroendocrine regulatory functions. In: E. J. Masoro & S. N Austad (Eds.), *Handbook of the Biology of Aging*, 5th ed. (pp. 297–323), San Diego: Academic Press.

Bartke, A., Brown-Borg, H., Mattison, J., Kinney, B., Hauck, S., & Wright, C. (2001a). Prolonged longevity of hypopituitary dwarf mice. *Exp. Gerontol. 36*: 21–28.

Bartke, A., Coschigano, K., Kophick, J., Chandrashekar, V., Mattison, J., Kinney, B., & Hauck, S. (2001b). Genes that prolong life: Relationship of growth hormone and growth to aging and life span. *J. Gerontol.: Biol. Sci. 56A:* B340–B349.

Bartke, A., Wright, J. C., Mattison, J. A., Ingram, D. K., Miller, R. A., & Roth, G. S. (2001c). Extending the lifespan of long-lived mice. *Nature 414*: 412.

Barzilai, N. & Gabriely, I. (2001). The role of fat depletion in the biological benefits of caloric restriction. *J. Nutrition 131*: 903S–906S.

Belage, M., Grizard, J., & Manin, M. (1990). Effect of calorie restriction on skeletal muscle and liver insulin binding in growing rats. *Horm. Metab. Res. 22*: 207–214.

Berg, B. N. & Simms, H. S. (1960). Nutrition and longevity in the rat. II. Longevity and the onset of disease with different levels of intake. *J. Nutrition 71:* 255–263.

Berg, T. F., Breen, P. J., Feuers, R. J., Oriaku, E. T., Chen, F. X., & Hart, R. W. (1994). Acute toxicity of ganciclovir: Effect of dietary restriction and chronobiology. *Fd. Chem. Toxic. 32*: 45–50.

Bergamini, E., Bombara, M., Fierabracci, V., Masiello, P., & Novelli, M. (1991). Effects of different regimens of dietary restriction on the age-related decline in insulin secretory response of isolated rat pancreatic islets. *Ann. N. Y. Acad. Sci. 621*: 327–336.

Birt, D. F., Yaktins, A., & Duysen, E. (1999). Glucocorticoid mediation of dietary energy restriction inhibition of mouse skin carcinogenesis. *J. Nutrition 129*: Suppl., 571S–586S.

Bodkin, N. L., Ortmeyer, H. K., & Hansen, B. C. (1995). Long-term dietary restriction in older-aged rhesus monkeys: Effects on insulin resistance. *J. Gerontol.: Biol. Sci. 50A*: B142–B147.

Boss, O., Samec, S., Kuhne, F., Bijlenga, P., Assimacopoulas-Jeannet, F., Seydoux, J., Giacobino, J-P., & Muzzin, P. (1998). Uncoupling Protein-3 expression in rodent skeletal muscle is modulated by food intake but not by changes in environmental temperature. *J. Biol. Chem. 273*: 5–8.

Breese, C. R., Ingram, R. L., & Sonntag, W. E. (1991). Influence of age and long-term dietary restriction on plasma insulin-like growth factor (IGF-1), IGF-1 gene expression, and IGF-1 binding proteins. *J. Gerontol.: Biol. Sci. 46*: B180–B187.

Brown-Borg, H. M., Borg, K. E., Meliska, C. J., & Bartke, A. (1996). Dwarf mice and the aging process. *Nature 384*: 33.

Cartee, G. D. & Dean, D. J. (1994). Glucose transport with brief dietary restriction: heterogenous responses in muscles. *Am. J. Physiol. 266*: E-946–E952.

Cartee, G. D., Kietzke, E. W., Briggs-Tung, C. (1994). Adaptation of muscle glucose transport with caloric restriction in adult, middle-aged, and old rats. *Am. J. Physiol. 266*: R1443–R1447.

Cefalu, W. R., Bell-Farrow, A. D., Wang, Z. Q., Sonntag, W. E., Fu, M-X, Baynes, J. W., & Thorpe, S. R. (1995). Caloric restriction decreases age-dependent accumulation of the glycoxidation products, N^ε-(carboxymethyl)lysine and pentosidine, in rat skin collagen. *J. Gerontol.: Biol. Sci. 50A*: B337–B341.

Cefalu, W. R., Wagner, J. D., Wang, Z. Q., Bell-Farrow, A. D., Collins, J., Haskell, D., Bechtold, R., Morgan, T. (1997). A study of caloric restriction and cardiovascular aging in cynomolgus monkeys (*Macaca fasicularis*): A potential model for aging research. *J. Gerontol.: Biol. Sci. 52A*: B10–B19.

Cerami, A. (1985). Hypothesis: Glucose as a mediator of aging. *J. Am. Geriatr. Soc. 33*: 626–634.

Chen, L. H. & Snyder, D. L. (1992). Effect of age, dietary restriction and germ-free environment on glutathione-related enzymes in Lobund-Wistar rats. *Arch. Gerontol. Geriatr. 14*: 17–26.

Chen, L. H., Saxon-Kelley, D. M., & Snyder, D. L. (1996) Effects of age and dietary restriction on liver endogenous antioxidant defenses in male Lobund-Wistar rats. *Age 19*: 101–109.

Chipalkatti, S., De, A, K., & Aiyar, A. S. (1983). Effect of diet restriction on some biochemical parameters and aging in mice. *J. Nutrition 113*: 944–950.

Choi, J. H. & Yu, B. P. (1994). Brain synaptosomal aging: free radicals and membrane fluidity. *Free Radic. Biol. Med. 18*: 133–139.

Chung, H-Y. & Yu, B. P. (2000). Significance of hepatic xanthine oxidase and uric acid in aged and dietary restricted rats. *Age 23*: 123–128.

Chung, H-Y., Kim, H-J., Shim, K-H., & Kim, K-W. (1999). Dietary modulation of prostanoid synthesis in the aging process: role of cyclooxygenase-2. *Mech. Ageing Dev. 111*: 97–106.

Chung, M. H., Kim, H-J., Nishimura, S., & Yu, B. P. (1992). Protection of DNA damage by dietary restriction. *Free Radic. Biol. Med. 12*: 523–525.

Clancy, D. J., Gems, D., Hafen, E., Leevers, S. J., & Partridge, L. (2002). Dietary restriction in long-lived flies. *Science 296*: 319.

Clancy, D. J., Gems, D., Harshman, L. G., Oldham, S., Stocker, H., Hafen, E., Leevers, S. J., & Partridge, L. (2001). Extension of life-span by loss of CHICO, a *Drosophila* insulin receptor substrate protein. *Science 292*: 104–106.

Cocchi, D., Cattaneo, L., Lane, M. A., Ingram, D. K., Cutler, R. G., & Roth, G. S. (1995). Effect of long-term dietary restriction on the somatotrophic axis in adult and aged monkeys. *Neuroendocrin. Lett. 17*: 181–186.

Cook, C. J. & Yu, B. P. (1998). Iron accumulation in aging: Modulation by dietary restriction. *Mech. Ageing Dev. 102*: 1–13.

Coschigano, K. T., Clemmons, D., Beooush, L. L, & Kopchick, J. J. (2000). Assessment of growth parameters and life span of GHR/BP gene-disrupted mice. *Endocrinology 141*: 2608–2613.

Craig, B. W., Garthwaite, S. M., & Holloszy, J. O. (1987). Adipocyte insulin resistance: Effects of aging, obesity, exercise, and food restriction. *J. Appl. Physiol. 62*: 95–100.

Davidson, M. (1978). Primary insulin antagonism of glucose transport in muscles from older-obese rat. *Metabolism 27*: 1994–2005.

Davidson, R. T., Arias, E. B., & Cartee, G. D. (2002). Caloric restriction increases muscle insulin action but not IRS-1-, IRS-2-, or phosphotyrosine- PI 3-kinase. *Am. J. Physiol. 282*: E270–E276.

Davis, L. J., Tadolini, B., Biagi, P. L., Walford, R. L., & Licastro, F. (1993). Effect of age and extent of dietary restriction on hepatic microsomal lipid peroxidation potential in mice. *Mech. Ageing Dev. 72*: 155–163.

Dean, D. J., & Cartee, G. D. (1996). Brief dietary restriction increases skeletal muscle glucose transport in old Fischer 344 rats. *J. Gerontol.: Biol. Sci. 51A*: B208–B213.

Dean, D. J., Brozinick, jr., J. T., Cushman, S. W., & Cartee, G. D. (1998a). Calorie restriction increases cell surface GLUT-4 in insulin-stimulated skeletal muscle. *Am. J. Physiol. 275*: E957–E964.

Dean, D. J., Gazdag, A. C., Wetter, T. J., & Cartee, G. D. (1998b). Comparison of the effects of 20 days and 15 months of calorie restriction in male Fischer 344 rats. *Aging Clin. Exp. Res. 10*: 303–307.

D'Costa, A. P., Lenham, J. E., Ingram, R. L., & Sonntag, W. E. (1993). Moderate caloric restriction increases type 1 IGF receptors and protein synthesis in aging rats. *Mech. Ageing Dev. 71*: 59–71.

Dhahbi, J. M., Mote, P. L., Wingo, J., Tillman, J. B., Walford, R. L., & Spindler, S. R. (1999). Calories and aging alter gene expression for gluconeogenic, glycolytic, and nitrogen-metabolizing enzymes. *Am. J. Physiol. 277*: E352–E360.

Dhahbi, J. M., Tillman, J. B., Cao, S., Mote, P. L., Walford, R. L., & Spindler, S. R. (1998). Caloric intake alters the efficiency of catalase mRNA translation in the liver of old female mice. *J. Gerontol.: Biol. Sci. 53A*: B180–B185.

Dubey, A., Forster, M. J., Lal, H., & Sohal, R. S. (1996). Effect of age and caloric intake on protein oxidation in different brain regions and on behavioral functions of the mouse. *Arch. Biochem. Biophys. 333*: 189–197.

Ducimetiere, P., Eschwege, E., Papoz, J. L., Richard, J. L., Claude, J. R., & Rosselin, G. (1980). Relationship of plasma insulin levels to the incidence of myocardial infarction and coronary heart disease mortality in middle-aged population. *Diabetologia 19*: 205–210.

Duffy, P. H., Feuers, R. J., Pipkin, J. L., Berg, T. F., Leakey, J. E. A., Turturro, A., & Hart, R. W. (1995). The effect of dietary restriction and aging on physiological response to drugs. In: R. W. Hart, D. A. Neuman, & R. T. Robertson (Eds.), *Dietary restriction: Implications for the design and interpretation of toxicity and carcinogenicity studies* (pp. 125–140). Washington, DC: ILSI Press.

Dunn, J. A., Patrick, J. S., Thorpe, S. R., & Baynes, J. W. (1989). Oxidation of glycated proteins: age-dependent accumulation of N-(carboxymethyl)-lysine in lens protein. *Biochemistry 28*: 9464–9468.

Dunn, S. E., Kari, F. W., French, J. R., Travlos, G., Wilson, R., & Barrett, J. C. (1997). Dietary restriction reduces insulin-like growth factor I levels, which modulates apoptosis, cell proliferation, and tumor progression in p[53]deficient mice. *Cancer Res. 57*: 4667–4672.

Enesco, H. E. & Kruk, P. (1981). Dietary restriction reduces fluorescent pigment accumulation in mice. *Exp. Gerontol. 16*: 357–361.

Escriva, F., Rodriguez, C., Cacho, J., Alvarez, C., Portha, B., & Pascual-Leone, A. M. (1992). Glucose utilization and insulin action in adult rats submitted to prolonged food restriction. *Am. J. Physiol. 263*: E1–E7.

Fabrizio, P., Pozza, F., Pletcher, S. D., Gendron, C. M, & Longo, V. D. (2001). Regulation of longevity and stress resistance by Sch9 in yeast. *Science 292*: 288–290.

Ferari, P. & Weidmann, P. (1990). Insulin, insulin sensitivity, and hypertension. *J. Hypertension 8*: 491–500.

Feuers, R. J., Duffy, P. H., Chen, F., Desai, V., Oriaku, E., Shaddock, J. G., Pipkin, J. W., Weindruch, R., & Hart, R. W. (1995). Intermediary metabolism and antioxidant systems. In: R. W. Hart, D. A. Neumann, & R. T. Robertson (Eds.), *Dietary Restriction: Implications for the Design and Interpretation of Toxicity and Carcinogenecity Studies*, (pp 181–195). Washington, DC: ILSI Press.

Feuers, R. J., Weindruch, R., & Hart, R. W. (1993). Caloric restriction, aging, and antioxidant enzymes. *Mutation Res. 295*: 191–200.

Feuers, R. J., Weindruch, R., Leakey, J. E. A., Duffy, P. H., & Hart, R. W. (1997). Increased effective activity of rat liver catalase by dietary restriction. *Age 20*: 215–220.

Finch, C. E. (1990). *Longevity, senescence, and the genome.* Chicago: University of Chicago Press.

Forster, M. J., Sohal, B. H., & Sohal, R. S. (2000). Reversible effects of long-term caloric restriction on protein oxidative damage. *J. Gerontol.: Biol. Sci. 55A*: B522–B529.

Flurkey, K., Papaconstantinou, J., & Harrison, D. E. (2002). The Snell dwarf mutation *Pit1[dw]* can increase life span in mice. *Mech. Ageing Dev. 123*: 121–130.

Flurkey, K., Papaconstaninou, J., Miller, R. A., & Harrison, D. E. (2001). Lifespan extension and delayed immune and collagen aging in mutant mice with defects in growth hormone production. *Proc. Natl. Acad. Sci. USA 98*: 6736–6741.

Furst, A. (1987). Hormetic effects in pharmacology: Pharmacological inversions as prototypes for hormesis. *Health Phys. 52*: 527–530.

Gazdag, A. C., Dumke, C. L., Kahn, C. R., & Cartee G. D. (1999). Calorie restriction increases insulin-stimulated glucose transport in skeletal muscle from IRS-1 knockout mice. *Diabetes 48*: 1930–1936.

Gazdag, A. C., Sullivan, S., Kemnitz, J. W., & Cartee, G. D. (2000). Effect of long-term caloric restriction on GLUT-4, phosphatidylinositol-3 kinase p85 subunit, and insulin receptor substrate-1 protein levels in rhesus monkey skeletal muscle. *J. Gerontol.: Biol. Sci. 55A*: B44–B46.

Girard, N., Ferland, G., Boulanger, L., & Gaudreau, P. (1998). Long-term calorie restriction protects rat pituitary growth hormone-releasing hormone binding sites from age-related alterations. *Neuroendocrinology 68*: 21–29.

Gong, X., Shang, F., Obin, M., Palmer, H., Scofrano, M. M., Jahngen-Hodge, J., Smith, D. E., & Taylor, A. (1997). Antioxidant enzyme activities in lens, liver, and kidney of calorie restricted Emory mice. *Mech. Ageing Dev. 99*: 181–192.

Gredilla, R., Sanz, A., Lopez-Torres, M., & Barja, G. (2001). Caloric restriction decreases mitochondrial free radical generation at complex I and lowers oxidative damage to mitochondrial DNA, *FASEB J 15*: 1589–1591.

Hamilton, M. L., Van Remmen, H., Drake, J. A., Yang, H., Guo, Z. M., Kewitt, K., Walter, C. A., & Richardson, A. (2001). Does oxidative damage to DNA increase with age? *Proc. Natl. Acad. Sci. USA 98*: 10469–10474.

Han, E. S., Evans, T. R., & Nelson, J. F. (1998). Adrenocortical responsiveness to adrenocorticotropic hormone is enhanced in chronically food restricted rats. *J. Nutrition 128*: 1415–1420.

Han, E-S., Evans, T. R., Shu, J. H., Lee, S., & Nelson, J. F. (2001). Food restriction enhances endogenous and corticotropin-induced plasma elevations of free but not total corticosterone throughout life in rats. *J. Gerontol.: Biol. Sci. 56A*: B391–B397.

Han, E-S., Levin, N., Bengani, N., Roberts, J. L., Suh, Y., Karelus, K., & Nelson, J. F. (1995). Hyperadrenocorticism and food restriction-induced life extension in the rat: Evidence for divergent regulation of pituitary proopiomelanocortin RNA and adrenocorticotropic hormone biosynthesis. *J. Gerontol.: Biol. Sci. 50A*: B288–B294.

Harris, S. B., Gunion, M. W., Rosenthal, M. J., & Walford, R. L. (1994). Serum glucose, glucose tolerance, corticosterone, and free fatty acids during aging in energy restricted mice. *Mech. Ageing Dev. 73*: 209–221.

Heydari, A. R., & Richardson, A. (1992). Does gene expression play any role in the mechanism of the antiaging effect of dietary restriction? *Ann. NY Acad. Sci. 663*: 384–395.

Heydari, A. R., Conrad, C. C., & Richardson, A. (1995). Expression of heat shock genes in hepatocytes is affected by age and food restriction in rats. *J. Nutrition 125*: 410–418.

Heydari, A. R., Wu, B., Takahashi, R., Strong, R., & Richardson, A. (1993). Expression of heat shock protein 70 is altered by age and diet at the level of transcription. *Mol. Cell. Biol. 13*: 2909–2918.

Imre, S. & Juhaz, E. (1987). The effect of oxidative stress on inbred mice of different ages. *Mech. Ageing Dev. 38*: 259–266.

Ivy, J. L., Young, J. C., Craig, B. W., Kobert, W. M., & Holloszy, J. O. (1991). Aging, exercise, and food restriction: Effect on skeletal muscle glucose uptake. *Mech. Ageing Dev. 61*: 123–133.

Ji, L. L., Dillon, D., & Wu, E. (1990). Alteration of antioxidant enzymes with aging in rat skeletal muscle and liver. *Am. J. Physiol. 258*: R918–R923.

Johnson, T. E., Cypser, J., de Castro, E., de Castro, S., Henderson, S., Murakami, S., Rikke, B., Tedesco, P., & Link, C. (2000). Gerontogenes mediate health and longevity in nematodes through increasing resistance to environmental toxins and stressors. *Exp. Gerontol. 35*: 687–694.

Johnson, T. E., Lithgow, G. J., & Murakami, S. (1996). Hypothesis: Interventions that increase the response to stress offer the potential for effective life prolongation and increased health. *J. Gerontol.: Biol. Sci. 51A*: B392–B395.

Kalant, N., Stewart, J., & Kaplan, R. (1988). Effect of diet restriction on glucose uptake. *Mech. Ageing Dev. 46*: 89–104.

Kalu, D. N., Orhii, P. B., Chen, C., Lee, D-Y., Hubbard, G. B., Lee, S., & Olatunji-Bello, Y. (1998). Aged-rodent models of long-term growth hormone therapy: Lack of deleterious effect on longevity. *J. Gerontol.: Biol. Sci. 53A*: B452–B463.

Kapahi, P., Boulton, M. E., & Kirkwood, T. B. L. (1999). Positive correlation between mammalian life span and cellular resistance to stress. *Free Radic. Biol. Med. 26*: 495–500.

Keenan, K. P., Ballam, G. C., Dixit, R., Soper, K. A., Laroque, P., Mattson, B. A., Adams, S. P., & Coleman, J. B. (1997). The effect of diet, overfeeding, and moderate dietary restriction on Sprague-Dawley rat survival, disease, and toxicology. *J. Nutrition 127*: Suppl., 851S–856S.

Kemnitz, J. W., Roecker, E. B., Weindruch, R., Elson, D. F., Baum, S. T., & Bergmann, R. N. (1994). Dietary restriction increases insulin sensitivity and lowers blood glucose in rhesus monkeys. *Am. J. Physiol. 266*: E540–E547.

Kim, J. D., Yu, B. P., McCarter, R. J. M., Lee, S. Y., & Herlihy, J. T. (1995). Exercise and diet modulate cardiac lipid peroxidation and antioxidant defenses. *Free Radic. Biol. Med. 20*: 83–88.

Kimura, K. D., Tissenbaum, H. A., Liu, Y., & Ruvkun, G. (1997). *daf-2*, an insulin receptor-like gene that regulates longevity and diapause in *Caenorhabditis elegans. Science 277*: 942–946.

Klebanov, S., Shehab, D., Stavinoha, W. B., Yongman, S., & Nelson, J. F. (1995). Hyperadrenocorticism attenuated inflammation, and the life prolonging action of food restriction in mice. *J. Gerontol.: Biol. Sci. 50A*: B78–B82.

Koizumi, A., Wada, Y., Tsukada, M., Hasegawa, J., & Walford, R. L. (1989). Low blood glucose levels and small islets of Langerhans in the pancreas of calorie-restricted mice. *Age 12*: 93–96.

Koizumi, A., Weindruch, R. & Walford, R. L. (1987). Influence of dietary restriction and age on liver enzyme activities and lipid peroxidation in mice. *J. Nutrition 117*: 361–367.

Kristal, B. S. & Yu, B. P. (1992). An emerging hypothesis: synergistic induction of aging by free radicals and Maillard reactions. *J. Gerontol.: Biol. Sci. 47*: B107–B114.

Kristal, B. S. & Yu, B. P. (1998). Dietary restriction augments resistance to oxidant-mediated inhibition of mitochondrial transcription. *Age 21*: 1–6.

Laganiere, S., & Yu, B. P. (1987). Anti-lipoperoxidation action of food restriction. *Biochem. Biophys. Res. Commun. 145*: 1185–1191.

Laganiere, S. & Yu, B. P. (1989). Effect of chronic food restriction in aging rats, II. Liver cytosolic antioxidants and related enzymes. *Mech. Ageing Dev. 48*: 221–230.

Lal, S. B., Ramsey, J. J., Monemdjou, S., Weindruch, R., & Harper, M-E. (2001). Effects of caloric restriction on skeletal muscle mitochondrial proton leak in aging rats. *J. Gerontol.: Biol. Sci. 56A*: B116–B122.

Lane, M. A. (2000). Metabolic mechanisms of longevity: Caloric restriction in mammals and longevity mutations in *Caenorhabditis elegans*; a common pathway? *J. Am. Aging Assoc. 23*: 1–7.

Lane, M. A., Ball, S. S., Ingram, D. K., Cutler, R. G., Engel, J., Read, V., & Roth, G. S. (1995). Diet restriction in rhesus monkeys lowers fasting and glucose-stimulated glucoregulatory end points. *Am. J. Physiol. 268*: E941–E948.

Lane, M. A., Tilmont, E. M., De Angelis, H., Handy, A., Ingram, D. K., Kemnitz, J. W., & Roth, G. S. (1999). Short-term calorie restriction improves disease-related markers in older rhesus monkeys (*Macaca mulatta*). *Mech. Ageing Dev. 112*: 185–196.

Lass, A., Sohal, B. H., Weindruch, R., Forster, M. J., & Sohal, R. S. (1998). Caloric restriction prevents age-associated accrual of oxidative damage to mouse skeletal muscle mitochondria. *Free Radic. Biol. Med. 25*: 1089–1097.

Leakey, J. E., Chen, S., Manjgaladze, M., Turturro, A., Duffy, P. H., Pipkin, J. L., & Hart, R. W. (1994). Role of glucocorticoids and "caloric stress" in modulating the effects of caloric restriction in rodents. *Ann. NY Acad. Sci. 719*: 171–194.

Lee, D. W. & Yu, B. P. (1990). Modulation of free radicals and superoxide dismutase by age and dietary restriction. *Aging Clin Exp. Res. 2*: 357–362.

Leeuwenburgh, C., Wagner, P., Holloszy, J. O., Sohal, R. S., & Heinecke, J. W. (1997). Caloric restriction attenuates dityrosine cross-linking of cardiac and skeletal muscle proteins in aging mice. *Arch. Biochem. Biophys. 346*: 74–80.

Lev-Ran, A. (1998). Mitogenic factors accelerate later-age disease: Insulin as a paradigm. *Mech. Ageing Dev. 102*: 95–113.

Li, Y., Yan, Q., Pendergrass, W. R., & Wolf, N. (1998). Response of lens epithelial cells to hydrogen peroxide stress and protective effect of caloric restriction. *Exp. Cell Res. 239*: 254–263.

Lin, K., Dorman, J. B., Rodan, A., & Kenyon, C. (1997). *Daf-16*: An HNF-3/forkhead family member that can function to double the life-span of *Caenorhabditis elegans*. *Science 278*: 1319–1322.

Lin, S-J., Defossez, P-A., & Guarente, L. (2000). Requirement of NAD and *SIR2* for life-span extension by calorie restriction in *Saccharomyces cerevisae*. *Science 289*: 2126–2128.

Lin, Y-J., Seroude, L., & Benzer, S. (1998). Extended life-span and stress resistance in the *Drosophila* mutant *methuselah*. *Science 282*: 943–946.

Lithgow, G. J. & Walker, G. A. (2002). Stress resistance as a determinant of *C. elegans* lifespan. *Mech. Ageing Dev. 123*: 765–771.

Luhtalta, T. A., Roecker, E. B., Pugh, T., Feuers, R. J., & Weindruch, R. (1994). Dietary restriction attenuates age-related increases in rat skeletal muscle antioxidant enzyme activities. *J. Gerontol.: Biol. Sci. 49*: B231–B238.

Martin, G. M., Austad, S. N., & Johnson, T. E. (1996). Genetic analysis of ageing: Role of oxidative damage and environmental stresses. *Nature Genet. 13*: 25–34.

Masoro, E. J. (1996). Possible mechanisms underlying the antiaging actions of caloric restriction. *Toxicol. Path. 24*: 738–741.

Masoro, E. J. (1998). Hormesis and the antiaging action of dietary restriction. *Exp. Gerontol. 33*: 61–66.

Masoro, E. J. (2001). Body composition. In: G. L. Maddox, (Ed), *The Encyclopedia of Aging*, Vol. I, 3rd ed. (pp. 131–132). New York: Springer Publishing Co.

Masoro, E. J., Compton, C., Yu, B. P., & Bertrand, H. (1983). Temporal and compositional dietary restrictions modulate age-related changes in serum lipids. *J. Nutrition 113*: 880–885.

Masoro, E. J., Katz, M. S., & McMahan, C. A. (1989). Evidence for the glycation hypothesis of aging from the food-restricted rodent model. *J. Gerontol.: Biol. Sci. 44*: B20–B22.

Masoro, E. J., McCarter, R. J. M., Katz, M. S., & McMahan, C. A. (1992). Dietary restriction alters characteristics of glucose fuel use. *J. Gerontol.: Biol. Sci. 47*: B202–B208.

Matsuo, M., Gomi, F., Kuramoto, K., & Sagai, M. (1993). Food restriction suppresses an age-dependent increase in exhalation rate of pentane from rats: a longitudinal study. *J. Gerontol.: Biol. Sci. 48*: B133–B138.

Mattison, J. A., Wright, C., Bronson, R. T., Roth, G. S., Ingram, D. K., & Bartke, A. (2000). Studies of aging in Ames dwarf mice: Effects of caloric restriction. *J. Am. Aging Assoc. 23*: 9–16.

Maynard Smith, J. (1958). Prolongation of the life of *Drosophila subobscura* by brief exposure of adults to high temperature. *Nature 181*: 496–497.

McCarter, R. J. (2000). Caloric restriction, exercise, and aging. In: C. K. Sen, L. Packer, & O. Hanninen, (Eds.), *Handbook of Oxidants and Antioxidants in Exercise* (pp. 797–829). Amsterdam: Elsevier Science.

McCarter, R. J., & Palmer, J. (1992). Energy metabolism and aging: A lifelong study of Fischer 344 rats. *Am. J. Physiol. 263*: E448–E452.

McCarter, R. J., Masoro, E. J., & Mejia, W. J. (1994). Effects of dietary restriction on glucose utilization of skeletal muscle in Fischer 344 rats. *Gerontologist 34*: 356 (Abstract).

McCay, C., Crowell, M., & Maynard, L. (1935). The effect of retarded growth upon the length of life and upon ultimate size. *J. Nutrition 10*: 63–79.

Mendel, C. M. (1989). The free hormone hypothesis: A physiologically based mathematical model. *Endocrine Rev. 10*: 232–274.

Migliaccio, E., Giorgio, M., Mele, S., Pelicci, G., Reboldi, P., Pandolfi, P. P., Lanfrancone, L., & Pelicci, P. G. (1999). The p66[shc] adaptor protein controls oxidative stress response and life span in mammals. *Nature 402*: 309–313.

Miller, R. A. (1999). Kleemeier award lecture: are there genes for aging? *J. Gerontol.: Biol. Sci. 54A*: B297–B307.

Monnier, V. M. (1990). Minireview: nonenzymatic glycosylation, the Maillard reaction and the aging process. *J. Gerontol.: Biol. Sci. 45*: B105–B111.

Moore, S. A., Lopez, A., Richardson, A., & Pahlavani, M. S. (1998). Effect of age and dietary restriction on the expression of heat shock protein 70 in rat alveolar macrophages. *Mech. Ageing Dev. 104*: 59–73.

Mote, P. L., Grizzle, J. M., Walford, R. L., & Spindler, S. R. (1991). Influence of age and caloric restriction on expression of hepatic genes for xenobiotic and oxygen metabolizing enzymes in the mouse. *J. Gerontol.: Biol. Sci. 46*: B95–B100.

Munck, A., Guyre, P. M., & Holbrook, N. J. (1984). Physiological functions of glucocorticoids in stress and their relation to pharmacological actions. *Endocrine Rev. 5*: 25–44.

Mura, C. V., Gong, X., Taylor, A., Villalobos-Molina, R., & Scrofano, M. M. (1996). Effects of calorie restriction and aging on the expression of antioxidant enzymes and ubiquitin in liver of Emory mice. *Mech. Ageing Dev. 91*: 115–119.

Neafsey, P. J. (1990). Longevity hormesis. A review. *Mech. Ageing Dev. 51*: 1–31.

Olefsky, J. M. & Reaven, E. M. (1975). Effect of age and obesity on insulin binding to isolated adipocytes. *Endocrinology 96*: 1486–1498.

Oriaku, E. T., Chen, F., Desai, V. G., Pipkin, J. L., Shaddock, J. G., Weindruch, R., Hart, R. W., & Feuers, R. J. (1997). A circadian study of liver antioxidant enzyme systems of female F344 rats subjected to dietary restriction for 6 weeks. *Age 20*: 221–228.

Ortmeyer, H. K., Bodkin, N. L., & Hansen, B. C. (1994). Chronic calorie restriction alters glycogen metabolism in rhesus monkeys. *Obesity Res. 2*: 549–555.

Ortmeyer, H. K., Huang, L., Larner, J., & Hansen, B. C. (1998). Insulin unexpectedly increases the glucose 6-phosphate K_a of skeletal muscle glycogen synthetase in calorie-restricted monkeys. *J. Basic Clin. Physiol. Pharmacol. 9*: 309–323.

Oster, M. H., Fiedler, P. J., Levin, N., & Cronin, M. J. (1995). Adaptation of the growth hormone and insulin-like growth factor-I axis to chronic and severe calorie or protein malnutrition. *J. Clin. Invest. 95*: 2258–2265.

Papaconstantinou, J., Reisner, P. D., Liu, L., & Kuninger, D. T. (1996). Mechanisms of altered gene expression with aging. In: E. L. Schneider & J. W. Rowe, (Eds.). *Handbook of the Biology of Aging*, 4th ed. (pp. 150–183). San Diego: Academic Press.

Parr, T. (1996). Insulin exposure controls the rate of mammalian aging. *Mech. Ageing Dev. 88*: 75–82.

Pashko, L. L. & Schwartz, A. G. (1992). Reversal of food restriction induced inhibition of mouse skin tumor promotion by adrenalectomy. *Carcinogenesis 10*: 1925–1928.

Pieri, C., Falaca, M., Marcheselli, F., Moroni, F., Recchioni, R., Marmocchi, F., & Lipidi, C. (1992). Food restriction in female Wistar rats. I. Lipid peroxidation and antioxidant enzymes in liver. *Arch. Gerontol. Geriatr. 14*: 93–99.

Pipkin, J. L., Hinson, W. G., Feuers, R. J., Lyn-Cook, L. E., Burns, E. R., Duffy, P. H., Hart, R., & Casciano, D. A. (1994). The temporal relationship of synthesis and phosphorylation in stress proteins 70 and 90 in aged caloric restricted rats exposed to bleomycin. *Aging Clin. Exp. Res. 6*: 125–132.

Quigley, K., Goya, R., Nachreiner, R., & Meites, J. (1990). Effects of underfeeding and refeeding on GH and thyroid hormone secretion in young, middle-aged, and old rats. *Exp. Gerontol. 25*: 447–457.

Rao, G., Xia, E., Nadakavukaren, M. J., & Richardson, A. (1990). Effect of dietary restriction on the age-dependent changes in the expression of antioxidant enzymes in rat liver. *J. Nutrition 120*: 602–609.

Rao, R. H. (1995). Fasting glucose homeostasis in the adaptation to chronic nutritional deprivation in rats. *Am. J. Physiol. 268*: E873–E879.

Reaven, E. P., Wright, D., Mondon, C. E., Solomon, R., Ho, H., & Reaven, G. M. (1983). Effect of age and diet on insulin secretion and insulin action in the rat. *Diabetes 32*: 175–180.

Reaven, G. M. (1989). *Clinician's guide to non-insulin-dependent diabetes mellitus*. New York: Marcel Dekker.

Reaven, G. M. & Reaven, E. P. (1981). Prevention of age-related hypertriglyceridemia by caloric restriction and exercise training in the rat. *Metabolism 30*: 982–986.

Reiser, K. M. (1994). Influence of age and long-term dietary restriction on enzymatically mediated crosslinks and nonenzymatic glycation of collagen in mice. *J. Gerontol.: Biol. Sci. 49*: B71–B79.

Richardson, A. (1991). Changes in the expression of genes involved in protecting cells against stress and free radicals. *Aging Clin. Exp. Res. 3*: 403–405.

Rikans, L. E., Moore, D. R., & Snowden, C. D. (1991). Sex-dependent differences in the effect of aging on antioxidant defence mechanisms of rat liver. *Biochim. Biophys. Acta 1074*: 195–200.

Rojas, C., Cadenas, S., Perez-Campo, R., Lopez-Torres, M. Pamplona, R., Prat, J., & Barja, G. (1993). Relationship between lipid peroxidation, fatty acid composition, and ascorbic acid in liver during carbohydrate and caloric restriction in mice. *Arch. Biochem. Biophys. 306*: 59–64.

Rossetti, L., Giaccari, A., & DeFronzo, R. A. (1990). Glucose toxicity. *Diabetes Care 13*: 610–630.

Ruggeri, B. A., Klurfeld, D. M., Kritchevsky, D., & Furlanetto, R. W. (1989). Caloric restriction and 7,12 dimethylbenz(*a*)anthracene-induced mammary tumor growth in rats: Alterations in circulating insulin, insulin-like growth factors I and II, and epidermal growth factor. *Cancer Res. 49*: 4130–4134.

Sabatino, F., Masoro, E. J., McMahan, C. A., & Kuhn, R. W. (1991). An assessment of the role of the glucocorticoid system in aging processes and in the action of food restriction. *J. Gerontol.: Biol. Sci. 46*: B171–B179.

Sacher, G. A. (1977). Life table modifications and life prolongation. In: C. E., Finch, & L. Hayflick, (Eds.), *Handbook of the Biology of Aging* (pp. 582–638). New York: Van Nostrand Reinhold.

Schwartz, A. G. & Pashko, L. L. (1994). Role of adrenocortical steroids in mediating cancer-prevention and age-retarding effects of food restriction in laboratory rodents. *J. Gerontol.: Biol. Sci. 49*: B37–B41.

Sell, D. R. (1997). Ageing promotes the increase of early glycation Amadori product as assessed by ε-N-(2-furoylmethyl)-L-lysine (furosine) levels in rodent skin collagen The relationship to dietary restriction and glycoxidation. *Mech. Ageing Dev. 95*: 81–99.

Sell, D. R. & Monnier, V. M. (1997). Age-related association of tail tendon break time with tissue pentosidine in DBA/2 vs C57BL/6 mice: The effect of dietary restriction. *J. Gerontol.: Biol. Sci. 52A*: B277–B284.

Sell, D. R., Lane, M. A., Johnson, W. A., Masoro, E. J., Mock, O. B., Reiser, K. M., Fogarty, J. F., Cutler, R. G., Ingram, D. K., Roth, G. S., & Monnier, V. M. (1996). Longevity and genetic determination of collagen glycoxidation kinetics in mammalian senescence. *Proc. Natl. Acad. Sci. USA 93*: 485–490.

Semsei, I., Rao, G., & Richardson, A. (1989). Changes in the expression of superoxide dismutase and catalase as a function of age and dietary restriction. *Biochem. Biophys. Res. Commun. 164*: 620–625.

Shimokawa, I., Higami, Y., Okimoto, T., Tomita, M., & Ikeda, T. (1996). Effects of lifelong dietary restriction on somatotropes: Immunochemical and functional aspects. *J. Gerontol.: Biol. Sci. 51A*: B396–B402.

Shimokawa, I., Higami, Y., Okimoto, T., Tomita, M., & Ikeda, T. (1997a). Effect of somatostatin-28 on growth hormone response to growth hormone-releasing hormone – Impact of aging and lifelong dietary restriction. *Neuroendocrinology 65*: 369–376.

Shimokawa, I., Tomita, M., Higami, Y., Okimoto, T., Kawahara, T, & Ikeda, T. (1997b). Dietary restriction maintains the basal rate of somatotrope renewal in later life in male rats. *Age 20*: 169–174.

Shimokawa, I., Yanagihara, K., Higami, Y., Okimoto, T., Tomita, M., Ikeda, T., & Lee, S. (2000). Effects of aging and dietary restriction on mRNA levels of receptors for growth hormone-releasing hormone and somatostatin in the rat pituitary. *J. Gerontol.: Biol. Sci. 55A*: B274–B279.

Sohal, R. S. & Dubey, A. (1994). Mitochondrial oxidative damage, hydrogen peroxide release, and aging. *Free Radic. Biol. Med. 16*: 621–626.

Sohal, R. S. & Weindruch, R. (1996). Oxidative stress, caloric restriction, and aging. *Science 273*: 59–63.

Sohal, R. S., Agarwal, S., Candas, M., Forster, M., & Lal, H. (1994a) Effect of age and caloric restriction on DNA oxidative damage in different tissues of C57BL/6 mice. *Mech. Ageing Dev. 76*: 215–224.

Sohal, R. S., Ku, H-H., Agarwal, S., Forster, M. J., & Lal, H. (1994b). Oxidative damage, mitochondrial oxidant generation, and antioxidant defenses during aging and in response to food restriction in the mouse. *Mech. Ageing Dev. 74*: 121–133.

Sonntag, W. E., Lenham, J. E., & Ingram, R. L. (1992). Effects of aging and dietary restriction on tissue protein synthesis: Relationship to plasma insulin-like growth factor-1. *J. Gerontol.: Biol. Sci. 47*: B159–B162.

Sonntag, W. E., Lynch, C. D., Cefalu, W. T., Ingram, R. L., Bennett, S. A., Thorton, P. L., & Khan, A. S. (1999). Pleiotropic effects of growth hormone and insulin-like growth factor (IGF)-1 on biological aging: Inferences from moderate caloric-restricted animals. (1999). *J. Gerontol.: Biol. Sci. 54A*: B521–B538.

Sonntag, W. E., Xu, X., Ingram, R. L., & D'Costa, A. P. (1995). Moderate caloric restriction alters the subcellular distribution of somatostatin mRNA and increases growth hormone pulse amplitude in aged animals. *Neuroendocrinology 61*: 601–608.

Spindler, S., Grizzle, J. M., Walford, R. L., & Mote, P. L. (1991). Aging and restriction of dietary calories increase insulin receptor mRNA, and aging increases glucocorticoid receptor mRNA in the liver of female C3B10F$_1$ mice. *J. Gerontol.: Biol. Sci. 46*: B233–B237.

Stewart, J., Meaney, M. J., Aitken, D., Jensen, L., & Kalant, N. (1988). The effects of acute and life-long food restriction on basal and stress-induced serum corticosterone levels in young and aged rats. *Endocrinology 123*: 1934–1941.

Stout, R. W. (1990). Insulin and atheroma. *Diabetes Care 13*: 631–654.

Tatar, M., Khazaeli, A. A., & Curtsinger, J. W. (1997). Chaperoning extended life. *Nature 390*: 30.

Tatar, M. Kopelman, A., Epstein, D., Tu, M.-P., Yin, C.-M., & Garofalo, R. S. (2001). A mutant *Drosophila* insulin receptor homolog that extends life-span and impairs neuroendocrine function. *Science 292*: 107–110.

Tavernakis, N. & Driscoll, M. (2002). Caloric restriction and life span: a role for protein turnover? *Mech. Ageing Dev. 123*: 215–219.

Taylor, A., Lipman, R. D., Jahngen-Hodge, J., Palmer, V., Smith, D., Padhye, N., Dallal, G. E., Cyr, D. E., Laxman, E., Shepherd, D., Morrow, F., Salomon, R., Perrone, G., Asmundsson, G., Meydani, M., Blumberg, J., Mune, M., Harrison, D. E., Archer, J. R., & Shigenaga, M. (1995). Dietary calorie restriction in Emory mouse: effects on lifespan, eye lens cataract prevalence and progression, levels of ascorbate, glutathione, glucose, and glycohemaglobin, tail collagen breaktime, DNA and RNA oxidation, skin integrity, fecundity, and cancer. *Mech. Ageing Dev. 79*: 33–57.

Tissenbaum, H. A., & Guarente, L. (2001). Increased doseage of a *SIR-2* gene extends the lifespan in *Caenorhabditis elegans*. *Nature 410*: 227–230.

Tomita, M., Shimokawa, I., Higami, Y., Yanagihara-Outa, K., Kawahara, T., Tanaka, K., Ikeda, T., & Shindo, H. (2001). Modulation by dietary restriction in gene expression related to insulin-like growth factor-1 in rat muscle. *Aging Clin. Exp. Res. 13*: 273–281.

Turturro, A., Hass, B., & Hart, R. W. (1998). Hormesis—Implications for risk assessment caloric intake (body weight) as an example. *Hum. Exp. Toxicol. 17*: 454–459.

Vanfleteren, J. R. & Braeckman, B. D. (1999). Mechanisms of life span determination in *Caenorhabditis elegans*. *Neurobiol. Aging 20*: 487–502.

Van Liew, J. B., Davis, P. J., Davis, F. G., Bernardis, L. L., Deziel, M. R., Marinucci, L. M., & Kumar, D. (1993). Effects of aging, diet, and sex on plasma glucose, fructosamine, and lipid concentrations in barrier-raised Fischer 344 rats. *J. Gerontol.: Biol. Sci. 48*: B184–B190.

Walford, R. L., Harris, S. B., & Gunion, M. W. (1992). The calorically restricted low-fat nutrient-dense diet in Biosphere 2 significantly lowers blood glucose, total leukocyte count, cholesterol, and blood pressure. *Proc. Natl. Acad. Sci. USA 89*: 11533–11537.

Wang, Z. Q., Bell-Farrow, A. D., Sonntag, W., & Cefalu, W. T. (1997). Effect of age and caloric restriction on insulin receptor binding and glucose transporter levels in aging rats. *Exp. Gerontol. 32*: 671–684.

Wetter, T. G., Gazdag, A. C., Dean, D. J., & Cartee, G. D. (1999). Effect of calorie restriction on in vivo glucose metabolism by individual tissues in rats. *Am. J. Physiol. 276*: E728–E738.

Wolkow, C. A., Kimura, K. D., Lee, M-S., & Ruvkun, G. (2000). Regulation of *C. elegans* life-span by insulinlike signaling in the nervous system. *Science 290*: 147–150.

Xia, E., Rao, G., Van Remmen, H., Heydari, A., & Richardson, A. (1995). Activities of antioxidant enzymes in various tissues of male Fischer 344 rats are altered by food restriction. *J. Nutrition 125*: 195–201.

Xu, X. & Sonntag, W. E. (1996a). Moderate caloric restriction prevents the age-related decline in growth hormone receptor signal transduction. *J. Gerontol.: Biol. Sci. 51A*: B167–B174.

Xu, X. & Sonntag, W. E. (1996b). Growth hormone-induced nuclear translocation of stat-3 decreases with age: modulation by caloric restriction. *Am. J. Physiol. 271*: E902–E909.

Yakar, S., Liu, J-L., Stannard, B., Butler, A., Accili, D., Sauer, B., & Le Roith, D. (1999). Normal growth and development in the absence of hepatic insulin-like growth factor I. *Proc. Natl. Acad. Sci. USA 96*: 7234–7329.

Youngman, L. D. (1993). Protein restriction and caloric restriction (CR) compared: effects on DNA damage, carcinogenesis, oxidative damage. *Mut. Res. 295*: 165–169.

Youngman, L. D., Park, J-Y. K., & Ames, R. N. (1992). Protein oxidation associated aging is reduced by dietary restriction of protein or calories. *Proc. Natl. Acad. Sci. USA 89*: 9112–9116.

Yu, B. P. (1996). Aging and oxidative stress: Modulation by dietary restriction. *Free Radic. Biol. Med. 21*: 651–668.

Zainal, Y., Oberly, T. D., Allison, D. B., Szweda, L. I., & Weindruch, R. (2000). Caloric restriction of rhesus monkeys lowers oxidative damage in skeletal muscle. *FASEB J. 14*: 1825–1836.

Zhang, Y. & Herman, B. (2002). Ageing and apoptosis. *Mech. Ageing Dev. 123*: 245–260.

CHAPTER 7

Evolution of the anti-aging action of caloric restriction

Contents

Are the anti-aging and life-prolonging actions of CR adaptations resulting from natural selection? If the answer to the question is affirmative, then questions emerge regarding the evolutionary pathway involved and whether it is similar or different among species. As of now, there are not definitive answers to these questions but a start has been made. However, before discussing emerging ideas, it is first necessary to briefly review current concepts regarding the evolutionary biology of aging.

Evolutionary biology of aging

Most evolutionary biologists no longer hold the once popular belief that aging is an evolutionary adaptation with a genetic program similar to that coding for development (Rose, 1991). Although there are several reasons for rejecting this view, probably the most important one is that senescence decreases the evolutionary fitness of the individual (i.e., decreases the ability to generate progeny), and thus makes it unlikely that aging is directly selected for by natural selection.

During the past 50 years or so, the view has emerged that aging evolved because the force of natural selection decreases with increasing

RESEARCH PROFILES IN AGING
VOLUME 1 ISSN 1567-7184

chronological age; i.e., there is a progressive age-associated decrease in the ability of natural selection to eliminate those detrimental traits that are expressed only at advanced ages. Although Charlesworth (1994) has provided a mathematical model for the quantitative assessment of the relationship between calendar age and the force of natural selection, the following simple hypothetical example makes it obvious that the force of natural selection decreases with increasing calendar age, albeit in qualitative terms: Assume that 1000 members of each gender of a cohort of animals reach reproductive maturity at age X; at that age, almost all of the 2000 members will be generating progeny, but as the age of the cohort increases, members will die (in the wild, the rate at which they die depends primarily on how hostile the environment). So with increasing age, fewer and fewer of those in the cohort will be available to generate progeny and thus add their gene pool to the next generation. Therefore, individuals with a gene with detrimental effects expressed only late in life will generate about the same number of progeny during their lifetime as those in the cohort that do not have this gene. That is, before the detrimental effects are expressed, all, or almost all, of the cohort are dead due to predation and other environmental hazards. That being the case, the ability of natural selection to eliminate this detrimental gene from the population is negligible. However, if such individuals are provided a protected environment, many may live to an age at which the detrimental effects are expressed, and they will suffer these effects, a phenomenon referred to as senescence.

Genetic mechanisms compatible with this theory of the evolution of aging have been the subject of considerable thought. Many years ago, Medawar (1952) proposed what is referred to as the mutation–accumulation mechanism. He realized that most deleterious mutations in gametes will result in progeny that are defective during most of life and that natural selection will act to remove such mutations from the population. However, he thought it likely that the deleterious effects of a few of the mutated genes in gametes will not result in defects in progeny until late in life, and thus will not be eliminated by natural selection. He postulated that these genes will accumulate in the genome and will be expressed by individuals living in protected environments and thus the occurrence of senescence.

A few years later, Williams (1957) proposed a genetic mechanism that is referred to as antagonistic pleiotropy. He postulated there are genes that increase evolutionary fitness early in life and have deleterious actions in late life. Since the force of natural selection is great in early life and negligible in late life, such genes would be selected for. Therefore, in this indirect way, senescence would also be selected for; i.e., senescence is a by-product of natural selection.

Although there is some evidence for the involvement of both the mutation-accumulation mechanism and the antagonistic pleiotropy mechanism, the Disposable Soma Theory of Kirkwood (1977) is compatible with the current view of the evolution of senescence and, in addition, it provides a basis for the evolution of CR's anti-aging action. Kirkwood's premise is that the fundamental life role of all organisms is to utilize free energy in the environment (e.g., the fuels in food) to generate progeny. To do so requires energy for reproduction but also requires energy for maintenance of the body (which he calls the soma or somatic maintenance). He proposes that the force of natural selection acts to apportion the use of energy between reproduction and somatic maintenance in a fashion that tends to maximize evolutionary fitness (i.e., tends to maximize the individual's lifetime yield of progeny). As a consequence, less energy is used for somatic maintenance than is needed for indefinite survival and hence the occurrence of senescence. He further proposes that the apportionment of energy between reproduction and somatic maintenance varies according to the environment in which the species or subgroup of a species evolved. If the species evolves in a highly hazardous environment, the greatest yield of progeny can be achieved when relatively little energy is used for somatic maintenance, enabling high rates of reproduction at young ages. However, these organisms will exhibit rapid senescence if introduced into a protected environment free of the environmental hazards that lead to their death at young ages during the evolution of the species. The converse can be said for species that evolved in environments that are not very hazardous. It is postulated that such animals will have a modest rate of reproduction that continues past young adulthood, and will use much energy for somatic maintenance; thus such animals will exhibit a slow rate of senescence.

The female reproductive life span hypothesis

Harrison and Archer (1989) addressed the question of the evolution of the life-prolonging and anti-aging actions of CR. Their hypothesis is based on the fact that in a protected environment, the span of female reproductive life of most species of mammals is about half of the maximal life span. They propose that in nature, periods of low availability of food for species like mice and rats (as might occur during a drought) may well outlast their reproductive life span. However, individuals whose genomes respond to food deficiency by slowing the rate of reproductive aging might have the ability to reproduce when ample food again becomes available under conditions favorable for both a successful pregnancy and survival of progeny. Such individuals would have an enormous selective advantage and would, therefore, become the dominant genotype of the species or subgroup of the species. Presumably, general organismic aging would slow in parallel with the slowing of reproductive aging. This hypothesis led to the prediction that CR's anti-aging actions should be greatest in those species with the shortest reproductive life span.

Phelan and Austad (1989) challenged this hypothesis on the grounds that reproductive senescence is largely irrelevant to life in the wild, because post-reproductive individuals are rarely found in nature. Thus, they feel that the force of natural selection to extend reproductive life span is likely to be very weak.

The energy apportionment hypothesis

Holliday (1989) proposed a hypothesis of the evolution of CR's life-prolonging action in terms of the Disposable Soma Theory of Aging. The basis of the hypothesis is that in nature, the availability of food would vary greatly over time with an excess of dietary calories at times, and food shortage at other times, the latter leading to starvation or near starvation. Holliday points out that it is known that during periods of restricted caloric intake, laboratory rats and mice exhibit either a reduction or cessation of reproduction. (See Chapter 4 for a detailed discussion of CR and reproduction.) Moreover, when the old CR rats and mice are provided with an abundance of food, they are able to reproduce at ages much greater than those at which *ad libitum*-fed animals have ceased reproduction. Based on this information, Holliday points out that in nature, rats and mice stop breeding during periods of starvation or near starvation and resume breeding again when food becomes available. By diverting energy from reproduction, Holliday concludes that the mice and rats on the restricted energy intake utilize more energy for somatic maintenance than do animals ingesting an abundance of calories. As a result, such animals have increased longevity and are able to generate progeny at a time when food resources are favorable for survival of offspring, clearly an evolutionary adaptation. In the case of the laboratory rodent on a sustained CR regimen, this translates into a marked retardation of senescence and increase in longevity.

Masoro and Austad (1996) embraced and expanded the view of Holliday. They pointed out that in nature, animals experience two kinds of food shortages: predictable shortages (such as those due to seasonal changes in climate) and unpredictable short-term shortages. To deal with the former, animals have evolved well-known physiological responses, such as seasonal reproductive cessation, topor, and hibernation. Masoro and Austad propose that the anti-aging actions of CR evolved in response to unpredictable food shortages. Although they, too, view this evolutionary adaptation in terms of the Disposable Soma Theory of Aging, they emphasize that a major player is the diverting of energy to increase the ability to cope with environmental stressors. During unpredictable food shortages, there is a driving need to seek food, and thus an increase in inevitable confrontations with stressors. As discussed in the section on hormesis in Chapter 6, CR increases the ability of the laboratory rodent to cope with stressors. Indeed, there is extensive literature on hormesis and the retardation of aging in invertebrate species (Johnson et al., 1996). In considering key factors in the evolution of CR's anti-aging action in rodents, Totter (1985) focused on the reduced use of energy for reproduction and its increased use for physical activity. Hart and Turturro (1998) also concluded that the increased ability to compete for available food at times of food shortage is an important factor driving CR's anti-aging action. In CR rodents, enhanced functioning of the glucocorticoid system and the induction of stress genes are likely proximate factors underlying the increased ability to cope with stressors and, in the laboratory setting, the retardation of aging.

Masoro and Austad based several predictions on their hypothesis. One is that other moderate environmental stressors, such as heat, would be expected to retard aging; as discussed in Chapter 6, this has been found to be the case for some invertebrate species. Also disabling the glucocorticoid system should reduce the anti-aging action of CR, and this has found to be the case in the effect of CR on carcinogenesis (Schwartz & Pashko, 1994). Of course, some animal species or populations, depending on their dietary mode or environment, may rarely, if ever, experience unpredictable short-term food shortages, and the prediction is that such animals are not likely to exhibit the anti-aging effects of CR.

The hibernation-like hypothesis

Walford and Spindler (1997) proposed that the effects of CR observed in laboratory rodents are part of a spectrum of evolutionary adaptive responses that include hibernation. They point out that reductions in body temperature and plasma glucose and insulin levels are cardinal features of CR in laboratory rodents as well as hibernation in mammals in the wild. Increases in gluconeogeneis and hepatic carbamyl phosphate synthetase I activity (the enzyme that catalyzes the first step in urea formation) are also characteristic of both CR and hibernation (Feuers et al., 1993; Storey & Storey, 1990; Walford & Spindler, 1997), as is an increase in antioxidant defenses (Buzadzic et al., 1990; Chapter 3). An increase in protein synthesis is also observed under both conditions (Nelson, 1989; Chapters 3 and 4). In addition, as shown in Turkish hamsters by Lyman et al. (1981), hibernation leads to life extension. It is interesting to note that mammals do not hibernate because of low environmental temperature, but rather because of food scarcity (Nedergaard & Cannon, 1990; Storey & Storey, 1990). Walford and Spindler suggest that CR and hibernation are only two examples of a spectrum of evolutionary adaptations to food deprivation. They also point out that neuroendocrine mechanisms play a fundamental role in hibernation (Wang, 1998), and that CR's anti-aging action probably utilizes similar mechanisms, with the likely involvement of enkephalin peptides.

Computer modeling

In regard to hypotheses on the evolution of CR's anti-aging actions, a major problem is the difficulty of testing them empirically. Shanley and Kirkwood (2000) developed a mathematical model of the mouse life history to determine if a hypothesis makes sense in the rigorous, quantitative way required by the mathematical dimension of Darwinian natural selection. They used information on the feeding, breeding, and survival of mice in nature to build a theoretical model of a "virtual wild mouse." Into the complex computer program of this "virtual wild mouse," they also built in information on the effects of CR on laboratory mice. They then asked the program how would the virtual mouse apportion the use of energy: (1) when provided an abundant food supply throughout life; and (2) when challenged with periods of food shortage. They found that with a continuous abundance of food, the virtual mouse would utilize sufficient energy for maintenance to live out its natural life expectancy in the wild. However, if challenged by periods of food shortage, the virtual mouse would invest an increased amount of energy for maintenance and thus would have increased longevity, compared to that of the virtual mouse with a continuous supply of abundant food. These findings are clearly in accord with the Energy Apportionment Hypothesis.

Testing hypotheses

Kirk (2001) has made a start on the empirical testing of evolutionary hypotheses of CR's life-prolonging and anti-aging actions. He used original and published data on how 10 species of rotifers respond to CR. Most species, but not all, responded to CR by increasing mean and maximum life span, mortality rate doubling time, initial mortality rate, and reproductive life span. Among the species, there was no correlation between the strength of the response to CR and reproductive life span, the age of first reproduction or total reproduction. CR increased longevity in species that reduced reproduction when starved, while CR decreased longevity in species that continued to reproduce when starved. The results of this study are in accord with the Energy Apportionment Hypothesis. Hopefully, this pioneering study by Kirk will lead the way to further empirical testing of hypotheses on a spectrum of species.

References

Buzadzic B., Spasic, M., Saicic, Z. S., Radoiicic, R., Petrovic, V. M., & Halliwell, B. (1990). Antioxidant defenses in the ground squirrel *Citellus citellus* 2, the effect of hibernation. *Free Radic. Biol. Med. 9*: 407–413.

Charlesworth, B. (1994). *Evolution in Age-Structured Populations* 2nd ed, London: Cambridge University Press.

Feuers, R. J., Weindruch, R., & Hart, R. W. (1993). Caloric restriction, aging and antioxidant activity. *Mutation Res. 295*: 191–200.

Harrison, D. E. & Archer, J. R. (1989). Natural selection for extended longevity from food restriction. *Growth Dev. Aging 53*: 3.

Hart, R. W. & Turturro, A. (1998). Evolution and dietary restriction. *Exp. Gerontol. 33*: 53–60.

Holliday, R. (1989). Food, reproduction, and longevity: Is the extended lifespan of calorie-restricted animals an evolutionary adaptation? *BioEssays 10*: 125–127.

Johnson, T. E., Lithgow, G. J., & Murakami, S. (1996). Hormesis: Interventions that increase the response to stress offer the potential for effective life prolongation and increased health. *J. Gerontol.: Biol. Sci. 51A*: B392–B395.

Kirk, K. L. (2001). Dietary restriction and aging: Comparative tests of evolutionary hypotheses. *J. Gerontol.: Biol. Sci. 56A*: B123–B129.

Kirkwood, T. B. L. (1977). Evolution of ageing. *Nature 270*: 301–304.

Lyman, A., O'Brien, R. C., Greene, G. C., & Papafrangos, E. D. (1981). Hibernation and longevity in the Turkish hamster, *Mesocricetus brandti. Science 212*: 668–670.

Masoro, E. J. & Austad, S. N. (1996). The evolution of the antiaging action of dietary restriction. *J. Gerontol.: Biol. Sci. 51A*: B387–B391.

Medawar, P. B. (1952). *An Unsolved Problem of Biology*. Oxford: Oxford University Press.

Nedergaard, J. & Cannon, B. (1990). Mammalian hibernation. *Phil. Trans. R. Soc. Lond. [Biol.] 326*: 669–686.

Nelson, R. A. (1989). Nitrogen turnover and its conservation in hibernation. In: A. Malau & B. Canquilhem, (Eds.), *Living in the Cold* (pp. 299–307). London: John Libby Eurotext Ltd.

Phelan, J. P. & Austad, S. N. (1989). Natural selection, dietary restriction, and extended longevity. *Growth Dev. Aging 53*: 4–5.

Rose, M. R. (1991). *Evolutionary Biology of Aging*. New York: Oxford University Press.

Schwartz, A. G. & Pashko, L. L. (1994). Role of adrenocortical steroids in mediating cancer-prevention and age-retarding effects of food restriction in laboratory rodents. *J. Gerontol.: Biol. Sci. 49*: B37–B41.

Shanley, B. P. & Kirkwood, T. B. L. (2000). Calorie restriction and aging: a life history analysis. *Evolution 54*: 740–750.

Storey, K. B. & Storey, J. M. (1990). Metabolic rate depression and biochemical adaptation in anaerobiosis, hibernation, and estivation. *Quarterly Rev. Biol. 65*: 145–174.

Totter, J. R. (1985). Food restriction, ionizing radiation, and natural selection. *Mech. Ageing Dev. 30*: 261–271.

Walford, R. L. & Spindler, S. R. (1997). The response to caloric restriction in mammals shows features also common to hibernation: A cross-adaptation hypothesis. *J. Gerontol.: Biol. Sci. 52A*: B179–B183.

Wang, L. C. H. (1988). Mammalian hibernation: an escape from the cold. In: R. Gilles (Ed.), *Advances in Comparative Environmental Physiology* (pp. 1–46). Berlin: Springer-Verlag.

Williams, G. C. (1957). Pleiotropy, natural selection, and the evolution of senescence. *Evolution 11*: 398–411.

CHAPTER 8

Caloric restriction mimetics

Contents

Serious efforts have been made to develop interventions that mimic CR's effects on aging. Ideally, such a mimicry would reproduce all the actions of CR. In practice, interventions that produce one or more of CR's effects on physiological processes, pathological processes, or longevity are referred to as CR mimetics. Of course, interventions that increase mean and maximal life span, as well as mortality rate doubling time of a population, are likely to produce basic effects on aging similar to those of CR. Chemical agents, alterations of physical activity, and genetically engineered mice are three interventions that have been investigated as potential CR mimetics. In such studies, it is critical to rule out the possibility of the intervention of decrease in food intake, because if reducing food intake is how the intervention affects aging, then it is not a CR mimetic, but merely another way of producing CR. Unfortunately, those investigating CR mimetics sometimes fail to take this factor into account.

Studies on CR mimetics have two goals. One, such studies may yield information on mechanisms responsible for the anti-aging actions of CR. Two, the information derived from such studies may lead to the development of interventions that could modulate the aging processes in humans.

Chemical agents

There is extensive literature on chemical agents whose effects on aging or age-associated diseases are similar to those of CR (Roth et al., 2001).

RESEARCH PROFILES IN AGING
VOLUME 1 ISSN 1567-7184

However, many of the studies do not deal with the agent's effects on longevity; rather, they focus on functional changes that may mediate CR's anti-aging action, such as the reduction in plasma glucose and insulin levels or CR's effects on age-associated diseases, e.g., cancer and neurodegenerative disease. The chemical agents most studied as CR mimetics are briefly discussed below.

Knoll (1988a) reported that deprenyl (a drug used to treat Parkinson's disease in humans) causes a remarkable increase in longevity in old male rats. When Wistar-Logan rats were treated with 0.25 mg deprenyl per kg body-weight three times per week, starting at 24 months of age and continuing until spontaneous death, mean length of life increased by 30.5% (Knoll et al., 1989). Other studies have confirmed the longevity-promoting action of deprenyl in rats, although the magnitude of the effect varied greatly. Milgram et al. (1990) treated male F344 rats with 0.25 mg of deprenyl per kg every other day, starting at 23–25 months of age; this regimen increased mean length of life by only 16.5%. Although Milgram et al. did not measure food intake, they assumed that it was not decreased by deprenyl since the body weight of the rats was not affected. Kitani et al. (1993) treated male F344 rats with 0.25 mg per kg three times a week, starting at either 18 or 24 months of age; mean length of life increased 15% when treatment was started at 18 months of age, and 34% when started at 24 months of age. However, an increase in longevity in response to deprenyl has not been observed in all rat studies. Bickford et al. (1997) found no clear effect on longevity of male F344 rats when deprenyl treatment was started at 52 weeks of age, while Gallagher et al. (1998) found that it decreases longevity when started at 3 months of age in male Wistar rats.

Yen and Knoll (1992) reported a 35.7% increase in median length of life in male OFA1 mice treated with deprenyl starting at 12 months of age. Piantanelli et al. (1994) started deprenyl treatment at 22 months of age in Balb/c-nu mice, and found only a 4.2% increase in median length of life. Archer and Harrison (1996) studied the effect of deprenyl administration started at 26 months of age in male and female B6D2F1 mice, and they found it extended mean length of life by 8.6% in both genders; they also treated B6CBAF1 mice with deprenyl starting at 18 months of age and found it increased the mean length of life 6.8% in females and 5.6% in males. In addition, they measured food intake and found it was not affected by deprenyl in these mouse strains. On the other hand, Ingram et al. (1993) found that deprenyl treatment started at 18 months of age did not affect longevity in male C57BL/6J mice.

Deprenyl treatment started at 13 months of age increases longevity in female but not in male Syrian hamsters (Stoll et al., 1997). Also, the survival of relatively healthy 10- to 15-year-old dogs is increased by deprenyl treatment (Ruehl et al., 1997).

In addition, deprenyl was found to restore sexual activity and improve learning ability of male rats (Knoll, 1988a; Knoll et al., 1989). However, Archer and Harrison (1996) reported that deprenyl decreased the number of pups sired by old male mice. It remains to be determined whether the apparent discrepancy between these two studies results from a species difference in the reproductive response to deprenyl or from an increase in sexual activity not resulting in increased fecundity. Although deprenyl enhances cognitive function in old rats, it does not have beneficial effects on motor function (Bickford et al., 1997); nor does it improve motor function in aged mice (Ingram et al., 1993).

The mechanism underlying the effects of deprenyl on longevity is not known. Knoll (1988b) suggested that it acts by protecting against toxic free radicals, and there is evidence that it enhances the systems that protect the corpus striatum of rats from the damaging effects of reactive oxygen molecules (Carrillo et al., 1991; Clow et al., 1991; Cohen & Spina, 1989; Knoll, 1988a). Kitani et al. (2001) have recently reviewed their extensive research on the ability of deprenyl administration to promote these protective systems in rat brain. It should also be noted that the administration of deprenyl has a similar effect on mouse and dog brain (Carrillo et al., 1994, 1996). Kitani et al. (1996) proposed that the life-extending action of deprenyl results from the up-regulation of superoxide dismutase and catalase activities of catecholaminergic neurons, thereby protecting these neurons from oxidative stress. This hypothesis is in line with the findings of subsequent studies by this group, which show that deprenyl up-regulates superoxide dismutase and catalase only in selective brain regions (Kitani et al., 1999).

Deprenyl may be considered a CR mimetic in that, like CR, it increases longevity in rodents and protects them against oxidative stress. However, even in regard to these actions, it must be pointed out that it differs from CR, which is most effective in extending length of life when begun early in life, and is without effect when initiated late in life (see Chapter 2). In contrast, deprenyl treatment appears to be most effective when begun at advanced ages. CR broadly protects the cells of rodents from damage due to oxidative stress, while deprenyl appears to mimic this action only in selective regions of the nervous system.

As discussed in Chapter 6, many believe that protection against oxidative damage is the mechanism underlying the life-prolonging and anti-aging actions of CR. Based on this view, it would be expected that many antioxidant substances are likely to be CR mimetics. Indeed, there have been many studies on the effects of antioxidant compounds on longevity in mice and rats. Reviewing the results of studies published prior to 1995, Yu (1995) concluded that the effects of such compounds were neither reproducible nor robust. In a more recent publication, Lipman et al. (1998) reported a study in which, starting at 18 months of age, male C57BL/6 mice were fed diets supplemented with vitamin E, or glutathione, or vitamin E plus glutathione, or strawberry extract; none of these supplements increased longevity. However, Joseph et al. (1998) found that when the diets of F344 rats from 6 to 15 months of age were supplemented with strawberry or spinach extract or vitamin E, the age-associated deterioration of neural function, including cognitive behavior, was retarded. They found spinach extract to be the most effective and suggest that its anti-aging action is due to the presence of antioxidants. In a subsequent study, they fed diets supplemented with extracts of strawberry or blueberry or spinach to 19-month-old F344 rats for 8 weeks, and they found that each supplement reversed age-associated deterioration in neural function, including cognitive and motor behavioral deficits (Joseph et al., 1999). The anthocyanins are the most effective compounds in blueberries in providing antioxidant protection (Galli et al., 2002).

Because of their antioxidant functions, both α-lipoic acid and coenzyme Q10 have also been suggested as CR mimetics (Weindruch et al., 2001). Oral administration of α-lipoic acid for 15 days has been reported to improve long-term memory and alleviate age-associated NMDA receptor deficits in 20–23-month-old female NMRI mice (Stoll et al., 1993). The administration of α-lipoic acid for two weeks to 24–26-month-old male F344 rats decreases oxidative stress and damage and improves hepatic mitochondrial function (Hagen et al., 1998b). Feeding acetyl-L-carnitine to old rats reverses the age-associated decline in mitochondrial function (Hagen et al., 1998a). And, feeding a combination of α-lipoic acid and acetyl-L-carnitine is particularly effective in restoring memory and mitochondrial function in old rats (Liu et al., 2002). Research on the effects of the administration of coenzyme Q10, α-lipoic acid, acetyl-L-carnitine or combinations of them on the longevity of mice and rats is needed to determine whether or not these substances may serve as CR mimetics.

Two studies suggest that *N-tert*-alpha-phenylbutylnitrone (PBN) may be a CR mimetic. Edamatsu et al. (1995) reported that the intraperitoneal administration of PBN to an accelerated senescence mouse strain, starting at about 20 weeks of age, increased the median length of life from 42 weeks for untreated mice to 56 weeks. Saito et al. (1998) administered PBN in the drinking water of C57BL/6 mice starting at 24.5 months of age, and found that it increased mean length of life from 29.0 months in untreated mice to 30.1 months, and maximum life span from 31.7 to 33.3 months, without affecting body weight. PBN has also been found to reverse (1) the age-associated decrease in striatal muscarinic receptor sensitivity when given for 14 days to F344 rats as old as 24 months of age (Joseph et al., 1995); and (2) the age-associated loss of cerebellar noradrenergic receptor function when given for 14 days to F344 rats 21–22 months of age (Gould & Bickford, 1994). Although PBN does have antioxidant activity (Carney et al., 1991), it may also have other cellular actions. Treatment with a novel nitrone, referred to as CPI-1429, delays memory impairment and mortality in mice (Floyd et al., 2002).

Melatonin has, among other functions, antioxidant activity. Pierpaoli and Regelson (1994) reported that when melatonin is administered in the drinking water of 15-month-old female BALB/c mice during the dark phase of the light cycle, both median and maximum life spans are extended. Similar findings were seen in NZB female mice when melatonin administration was started at 5 months of age, and in C57BL/6 male mice when started at 19 months of age. Administration of melatonin to fruit flies was found to significantly increase both the median and maximum life span and to enhance the ability to cope with oxidative stress and heat stress (Bonilla et al., 2002). In contrast, Lipman et al. (1998) found that when the diet of male mice was supplemented with melatonin, starting at 18 months of age, there was no effect on median length of life.

The potential role of the glucose–insulin system in the anti-aging, life-prolonging actions of CR was discussed in Chapter 6. Not surprisingly, chemical agents that affect that system are viewed as CR mimetic candidates. Dilman and Anisimov (1980) studied the effect of phenformin (which lowers blood glucose levels and increases insulin sensitivity) on the longevity of C3H/Sn mice and found that there was a 23% increase in mean length of life. Since chromium picolate influences carbohydrate metabolism in a fashion similar to phenformin (McCarty, 1993), it is also a CR mimetic candidate. Indeed, Evans and Meyer (1992) published an abstract in which it was stated that chromium picolate markedly increases the median length of life of Long-Evans rats.

Lane et al. (1998) studied 2-deoxy-D-glucose (2DG) as a potential CR mimetic. 2DG, an analog of glucose, is phosphorylated to the 6-phosphate form in the first step of glucose metabolism via the Embden-Meyerhof pathway; but unlike glucose, it cannot be metabolized further.

Rather, the phosphorylated form accumulates, thereby inhibiting the metabolism of glucose. Lane et al. fed rats a diet containing 0.4% by weight 2DG and found that these rats exhibited physiological characteristics similar to rats on a CR regimen, including a decrease in plasma insulin, a modest decrease in plasma glucose, and a decrease in body temperature. There was no decrease in food intake, and body weight was the same or slightly less than that of the control rats. The rats on the 2DG diet began dying at about 12 months of age (Mattson et al., 2001b); the investigators felt this might be due to the 0.4% level in the diet being too high, and are now carrying out a longevity study using a diet with 2DG at the 0.25% level. 2DG has been found to inhibit neurodegeneration in a manner similar to that of CR in rodent models of neurodegenerative disease (Duan & Mattson, 1999; Guo & Mattson, 2000; Mattson et al., 2001a).

As discussed in Chapter 6, the glycation theory of aging views plasma glucose as a contributor to senescence because of its role in the glycation of proteins and DNA. Thus, according to this concept, the reduction in plasma glucose concentration by CR plays an important role in its anti-aging action. It is further believed that glycation promotes senescence because it leads to the formation of advanced glycation end-products (AGE), which are damaging to proteins and DNA as well as cellular functions. If this concept is valid, chemical agents that inhibit the formation of AGE are CR mimetic candidates. Aminoguanidine has been found to inhibit the formation of AGE (Brownlee, 1995), although, the chemical mechanisms involved are still the subject of investigation (Sell et al., 2001). Thus, aminoguanidine is also a potential CR mimetic. When administered to rats, aminoguanidine retards the age-associated progression of cardiovascular and renal pathology (Li et al., 1996). It also decreases the age-associated stiffening of arteries in male WAG/Rij rats (Corman et al., 1998). There are other chemical agents, such as thiazolium compounds and pyridoxamine, that inhibit glycation cross-linking (Bailey, 2001), and such compounds are also potential CR mimetics.

Long-term administration of dehydroepiandrosterone (DHEA), a steroid secreted by the adrenal cortex, suppresses the spontaneous occurrence of mammary tumors in C3H(Avy/a) female mice (Schwartz, 1979). Nelson (1995) has reviewed many of the studies done since 1979 that showed DHEA administration retards both spontaneous and carcinogen-induced cancers as well as age changes in neural and immune function. Since many of these actions also occur with CR, it would appear that DHEA administration is a CR mimetic. However, this conclusion must be

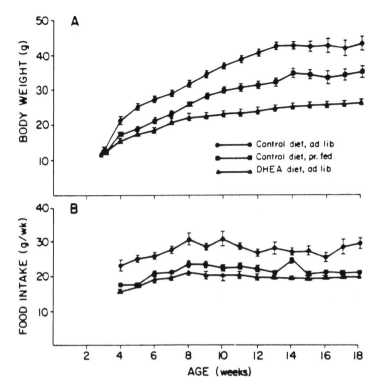

Figure 8-1. Body weight and food intake of mice fed a control diet *ad libitum* or a DHEA-containing diet *ad libitum* (0.4% of diet by weight) or a control diet pair fed to the intake of mice on the DHEA-containing diet. Body weight (Panel A) and food intake (Panel B). (From Weindruch et al., 1984.)

tempered by a study of Weindruch et al. (1984) who found that DHEA reduces food intake (Figure 8-1). Thus, rather than being a CR mimetic, DHEA may instead be just another way of reducing food intake.

In summary, many chemical agents have been found to have some effects that are similar to those of CR. However, as yet, much more needs to be done to establish any as a CR mimetic within the context of gerontology.

Alterations of physical activity

Currently, it is generally believed that exercise promotes healthy aging and possibly longevity as well (Lee et al., 1997; Rowe & Kahn, 1998). Interestingly, Holloszy and Kohrt (1995) point out that in the past, many gerontologists used the "rate of living" concept of aging to conclude that exercise would accelerate aging and decrease longevity. Indeed, early studies on rats were in accord with this belief, in that those on an exercise program died earlier than the rats kept sedentary (Benedict & Sherman, 1937; Slonaker, 1912). Based on their own experience, Holloszy and Kohrt feel it is likely that the findings of these early workers may have been due to infectious diseases, particularly lung infection, which was rampant in laboratory rat holding facilities prior to the development of technology for producing and maintaining specific pathogen-free (SPF) rats and mice.

Also, contrary to the early studies, several investigators have found that exercise extends the life of rats. Retzlaff et al. (1966) reported that moderate-intensity exercise increased both the average length of life and maximum life span of Sprague-Dawley rats. Edington et al. (1972) reported that daily treadmill exercise increased the longevity of young rats but reduced survival of old rats. Goodrick (1980) studied the effect of long-term voluntary wheel-running on the longevity of Wistar rats, and found that exercise started at 6 weeks of age increased both the mean and maximum length of life. However, when Goodrick et al. (1983) started this exercise program at 10.5–18 months of age, it had no effect on the longevity of the rats. The finding that exercise can increase both the mean and maximum length of life leads to conclusion that it is a CR mimetic.

However, research on Long-Evans rats conducted in Holloszy's laboratory does not fully support this conclusion. In their studies (Holloszy & Schlectman, 1991; Holloszy et al., 1985), 6-month-old male rats were divided into the following three groups; an *ad libitum*-fed sedentary group; an *ad libitum*-fed wheel-runner group; and a sedentary group whose food intake was restricted so as to maintain a body weight similar to the wheel-runner group (a so-called paired-weight group). Initially, the wheel-runner group voluntarily exercised vigorously; but after a few months, voluntary exercise decreased markedly. In order to maintain high levels of exercise, the amount of food given to the wheel-runner group had to be reduced to 92% of the *ad libitum*-fed sedentary group, and because of this, a fourth group of rats was needed, namely sedentary rats provided with 92% the amount of food consumed by the *ad libitum*-fed sedentary group. In regard to food intake, it was surprising to find that the *ad libitum*-fed wheel-runner group did not eat more than the *ad libitum*-fed sedentary group. Thus for the paired-weight group to maintain a body weight similar to that of the wheel-runners, they were fed about 70% of the intake of *ad libitum*-fed sedentary group.

The mean and maximum longevities were found to be the same for the *ad libitum*-fed sedentary group and the sedentary group fed 92% that much food. The wheel-runners exhibited an increased mean length of life but no increase in maximum length of life, while the sedentary paired-weight group, which was on a moderate CR regimen, exhibited a marked increase in both average and maximum length of life.

Unlike male Long-Evans rats, the females increase food intake when on an exercise program. Both the male and female rats respond to wheel-running exercise with an increase in mean length of life but not in maximum length of life (Holloszy, 1993). Comparing the longevity characteristics of F344 male rats that were either sedentary or wheel-runners, McCarter et al. (1997) found that wheel-running increased median length of life but did not influence maximum length of life (measured by the age of tenth percentile survivors). The increase in median length of life may be due to the fact that exercise delays the occurrence and progression of age-associated neoplastic and non-neoplastic diseases (Ikeno et al., 1997).

The studies of the Holloszy group and the McCarter group show that unlike CR, exercise does not increase the maximum length of life. Thus, unlike the findings of Goodrick and associates, those of the Holloszy and McCarter groups are not in accord with exercise as a CR mimetic.

Genetically engineered mice

The ability to increase, decrease, or eliminate the functioning of a specific mouse gene provides a powerful tool for the experimental generation of CR mimetic models. Such models have the potential of being invaluable for research on the basic mechanisms by which CR retards aging processes and extends longevity. While work in this area is only beginning, interesting models have already emerged.

The research of Migliaccio et al. (1999) provides an excellent example of such a model. Using the technique of targeted mutation, they produced mice in which the $p66^{shc}$ gene is ablated; these mice had a 30% increase in longevity as well as enhanced resistance to oxidative stress. Further work is needed to establish this mouse as a CR mimetic model; if it is, this model will play an important role in the investigation of the mechanisms responsible for CR's anti-aging actions.

The targeted disruption of the growth hormone receptor–growth hormone binding protein gene in mice is another promising example (Coschigano et al., 2000). These mice have an extended life span and, as discussed in Chapter 6, they have already yielded evidence to support a role of the growth hormone-IGF-1-axis in the anti-aging and life-prolonging actions of CR.

Transgenic mice that overexpress the Glut-4 transporter show a long-term reduction in plasma glucose, while plasma insulin levels remain the same as those of the wild-type mice (McCarter, personal communication). Such mice should prove invaluable for exploring the glycemia component of the Alteration of the Glucose–Insulin System hypothesis (see Chapter 6) of CR's anti-aging action. Indeed, McCarter and his colleagues are in the process of conducting such a study.

However, it is important to note two caveats concerning the use of genetically engineered mice in gerontologic research. One is the need to make certain that the genetic alteration does not have effects in addition to the expected one. The other problem stems from the fact that the gene undergoes alteration throughout the life of the mouse, and thus can affect development as well as aging. These problems cloud the gerontologic interpretation of studies based on genetically engineered mice. Fortunately, technology is being developed that enables the expression of genetic alterations in an age-dependent fashion.

References

Archer, J. R. & Harrison, D. E. (1996). L-deprenyl treatment in aged mice slightly increases life span, and greatly reduces fecundity by aged males. *J. Gerontol.: Biol. Sci.* *51A*: B448–B453.

Bailey, A. J. (2001). Molecular mechanisms of ageing in connective tissues. *Mech. Ageing Dev. 122*: 735–755.

Benedict, G. & Sherman, H. C. (1937). Basal metabolism of rats in relation to old age and exercise during old age. *J. Nutrition 14*: 179–198.

Bickford, P. C., Adams, C. E., Boyson, S. J., Curella, P., Gerhardt, G. A., Heron, C., Ivy, G. O., Lin, A. M. L. Y., Murphy, M. P., Poth, K., Wallace, D. R., Young, D. A., Zahniser, N. R. & Rose, G. M. (1997). Long-term treatment of male F344 rats with deprenyl: Assessment of effects on longevity, behavior, and brain function. *Neurobiol. Aging 18*: 309–318.

Bonilla, E., Medina-Leendertz, S., & Diaz. S. (2002). Extension of life span and stress resistance of *Drosophila melanogaster* by long-term supplementation with melatonin. *Exp. Gerontol. 37*: 629–638.

Brownlee, M. (1995). Advanced protein glycosylation in diabetes and aging. *Annu. Rev. Med. 46*: 223–234.

Carney, J. M., Starke-Reed, P., Oliver, C. N., Landrum, R. W., Cheng, M. S., & Weed, F. (1991). Reversal of age-related increase in brain protein oxidation, decrease in enzyme activity, and loss in temporal and spatial memory by chronic administration of the spin-trapping compound N-tert-butyl-alpha-phenylnitrone. *Proc. Natl. Acad. Sci. USA 88*: 3633–3636.

Carrillo, M. C., Ivy, G. O., Milgram, N. W., Head, E., Wu, P., & Kitani, K. (1994). Deprenyl increases activities of superoxide dismutase (SOD) in striatum of dog brain. *Life Sci. 54*: 1483–1489.

Carrillo, M. C., Kanai, S., Nobubo, M., & Kitani, K. (1991). (−) Deprenyl induces activities of both superoxide dismutase and catalase but not of glutathione peroxidase in the striatum of young male rats. *Life Sci. 48*: 517–521.

Carrillo, M. C., Kitani, K., Kanai, S., Sato, Y., Ivy, G. O., & Miyasaka, K. (1996). Long term treatment with (−)deprenyl reduces the optimal dose as well as the effective dose range for increasing antioxidant enzyme activities in old mouse brain. *Life Sci. 59*: 1047–1057.

Clow, A., Hussain, T., Glover, V., Sandler, M., Dexter, D. T., & Walker, M. (1991). (−)Deprenyl can induce soluble superoxide dismutase in rat striata. *J. Neural Transm. Gen. Sect. 86*: 77–80.

Cohen, G. & Spina, M. B. (1989). Deprenyl depresses the oxidative stress associated with increased dopamine turnover. *Ann. Neurol. 26*: 689–690.

Corman, B., Duriez, M., Poitevin, R., Heudes, D., Bruneval, P., Tedgui, A., & Levy, B. I. (1998). Aminoguanidine prevents age-related arterial stiffening and cardiac hypertrophy. *Proc. Natl. Acad. Sci. USA 95*: 1301–1306.

Coschigano, K. T., Clemmons, D., Beooush, L. L., & Kopchick, J. J. (2000). Assessment of growth parameters and life span of GHR/BP gene-disrupted mice. *Endocrinology 141*: 2608–2613.

Dilman, V. M. & Anisimov, V. N. (1980). Effect of treatment with phenformin, diphenylhydantoin or L-Dopa on life span and tumour incidence of C3H/Sn mice. *Gerontology 26*: 241–246.

Duan, W. & Mattson, M. P. (1999). Dietary restriction and 2-deoxyglucose administration improve behavioral outcome and reduce degeneration of dopaminergic neurons in models of Parkinson's disease. *J. Neurosci Res. 57*: 195–206.

Edamatsu, R., Mori, A., & Packer, L. (1995). The spin-trap N-tert-alpha-phenylbutylnitrone prolongs the life span of the senescence accelerated mouse. *Biochem. Biophys. Res. Comm. 211*: 847–849.

Edington, D. W., Cosmas, A. C., & McCaferty, W. B. (1972). Exercise and longevity: evidence for a threshold age. *J. Gerontol. 27*: 341–343.

Evans, G. W. & Meyer, L. (1992) Chromium picolate increases longevity. *Age 15*: 134 (abstract).

Floyd. R. A., Hensley, K., Forster, M. J., Kelleher-Anderson, J. A., & Wood, P. L. (2002). Nitrones as neuroprotective and antiaging drugs. *Ann NY Acad. Sci. 959*: 321–329.

Gallagher, I. M., Claw, A., & Glover, V. (1998). Long term administration of (−)-deprenyl increases mortality in male Wistar rats. *J. Neural Transm. Suppl. 52*: 315–320.

Galli, R. L., Shukitt-Hale, B., Youdim, K., & Joseph, J. A. (2002). Fruit polyphenolics and brain aging. Nutritional interventions targeting age-related neuronal and behavioral deficits. *Ann. NY Acad. Sci. 959*: 128–131.

Goodrick, C. L. (1980). Effects of long-term voluntary wheel exercise on male and female Wistar rats: 1. Longevity, body weight and metabolic rate. *Gerontology 26*: 22–23.

Goodrick, C. L., Ingram, D. K., Reynolds, M. A., Freeman, J. R., & Cider, N. L. (1983). Differential effects of intermittent feeding and voluntary exercise on body weight and lifespan in adult rats. *J. Gerontol. 38*: 36–45.

Gould, T. J. & Bickford, P. C. (1994). The chronic effects of treatment with N-tert-butyl-alpha-phenylnitrone in cerebellar noradrenergic receptor function in aged F344 rats. *Brain Res. 660*: 333–336.

Guo, Z. H. & Mattson, M. P. (2000). In vivo 2-deoxyglucose administration preserves glucose and glutamate transport and mitochondrial function in cortical synaptic terminals after exposure to amyloid-peptide and iron: Evidence for a stress response. *Exp. Neurol. 166*: 173–179.

Hagen, T. M., Ingersoll, R. T., Lykkesfeldt, J., Liu, J., Wehr, C. M. Vinarsky, V., Bartholomew, J. C., & Ames, B. (1998a). (R)-α–lipoic acid-supplemented old rats have improved mitochondrial function, decreased oxidative damage, and increased metabolic rate. *FASEB J. 13*: 411–418.

Hagen, T. M., Ingersoll, R. T., Wehr, C. M., Lykkesfeldt, J., Vinarsky, V., Bartholomew, J. C., Song, M-H., & Ames, B. (1998b). Acetyl-L-carnitine fed to old rats partially restores mitochondrial function and ambulatory activity. *Proc. Natl. Acad. Sci. USA 95*: 9562–9566.

Holloszy, J. O. (1993). Exercise increases average longevity of female rats despite increased food intake and no growth retardation. *J. Gerontol.: Biol. Sci. 48*: B97–B100.

Holloszy, J. O. & Kohrt, W. M. (1995). Exercise. In: E. J. Masoro (Ed.), *Handbook of Physiology*, Section 11, *Aging* (pp. 633–666). New York: Oxford University Press.

Holloszy, J. O. & Schlectman, K. B. (1991). Interactions between exercise and food restriction: effects on longevity of male rats. *J. Appl. Physiol. 70*: 1529–1535.

Holloszy, J. O., Smith, E. K., Vining, M., & Adams, S. (1985). Effect of voluntary exercise on longevity of rats. *J. Appl. Physiol. 59*: 826–831.

Ikeno, Y., Bertrand, H. A., & Herlihy, J. T. (1997). Effects of dietary restriction and exercise on the age-related pathology of the rat. *Age 20*: 107–118.

Ingram, D. K., Wiener, H. L., Chachich, M. E., Long, J. M., Hengemihle, J., & Gupta, M. (1993). Chronic treatment of aged mice with L-deprenyl produces marked striatal MAO-B inhibition but no beneficial effects on survival, motor performance, or nigral lipofuscin accumulation. *Neruobiol. Aging 14*: 431–440.

Joseph, J. A., Cao, G., & Cutler, R. C. (1995). In vivo and in vitro administration of the nitrone spin-trapping compound, N-tert-butyl-a-phenylnitrone (PBN), reduces age-related deficits in striatal muscarinic receptor sensitivity. *Brain Res. 671*: 73–77.

Joseph, J. A., Shukitt-Hale, B., Denisova, N. A., Bielinski, D., Martin, A., McEwen, J. J., & Bickford, P. C. (1999). Reversals of age-related declines in neuronal signal transduction, cognitive, and motor behavioral deficits with blueberry, spinach or strawberry supplementation. *J. Neurosci. 19*: 8114–8121.

Joseph, J. A., Shukitt-Hale, B., Denisova, N. A., Prior, R. L., Cao, G., Martin, A., Taglialatela, G., & Bickford, P. C. (1998). Long-term dietary strawberry, spinach, or vitamin E supplementation retards the onset of age-related signal-transduction and cognitive behavioral deficits. *J. Neurosci. 18*: 8047–8055.

Kitani, K., Kanai, S., Ivy, G. O., & Carrillo, M. C. (1999). Pharmacologic modifications of endogenous antioxidant enzymes with special reference to the effects of deprenyl: A possible antioxidant strategy. *Mech. Ageing Dev. 111*: 211–221.

Kitani, K., Kanai, S., Sato, Y., Ohta, M., Ivy, G. O., & Carrillo, M. C. (1993). Chronic treatment of (−) deprenyl prolongs the life span of male Fischer 344 rats. Further evidence. *Life Sci. 52*: 281–288.

Kitani, K., Minami, C., Yamamoto, T., Maruyama, W., Kanai, S., Ivy, G. O., & Carrillo, M. C. (2001). Do antioxidant strategies work against aging and age-associated disorder? Propargylamines: A possible antioxidant strategy. *Ann. NY Acad. Sci. 928*: 248–260.

Kitani, K., Miyasaka, K., Kanai, S., Carrillo, M. C., & Ivy, G. O. (1996). Upregulation of antioxidant enzyme activity by deprenyl. Implications for life span extension. *Ann. NY Acad. Sci. 786*: 391–409.

Knoll, J. (1988a). The striatal dependency of lifespan in male rats. Longevity study with (−) deprenyl. *Mech. Ageing Dev. 46*: 237–262.

Knoll, J. (1988b). (−) Deprenyl (selegine, movergan) facilitates the activity of nigrostriatal dopaminergic neuron. *J. Neural. Transm. Suppl. 25*: 45–66.

Knoll, J., Dallo, J., & Yen, T. T. (1989). Striatal dopamine, sexual activity and lifespan longevity of rats treated with (−)deprenyl. *Life Sci. 45*: 525–531.

Lane, M. A., Ingram, D. K., & Roth, G. S. (1998). 2-Deoxyglucose feeding to rats mimics physiologic effects of calorie restriction. *J. Anti-Aging Med. 1*: 327–337.

Lee, I.-M., Paffenbarger, jr., R. S., & Hennekens, C. H. (1997). Physical activity, physical fitness, and longevity. *Aging Clin. Exp. Res. 9*: 2–11.

Li, Y. M., Steffes, M., Donnelly, T., Liu, C., Fuh, H., Basgen, J., Bucala, R., & Vlassara, H. (1996). Prevention of cardiovascular and renal pathology of aging by the advanced glycation inhibitor aminoguanidine. *Proc. Natl. Acad. Sci. USA 93*: 3902–3907.

Lipman, R. D., Bronson, R. T., Wu, D., Smith, D. E., Prior, R., Cao, G., Han, S. N., Martin, K. R., Meydani, S. N., & Meydani, M. (1998). Disease incidence and longevity are unaltered by dietary antioxidant supplementation initiated during middle age in C57BL/6 mice. *Mech. Ageing Dev. 103*: 269–284.

Liu, J., Atamma. H., Kuratsume, H., & Ames, B. (2002). Delaying brain mitochondrial decay and aging with mitochondrial antioxidants and metabolites. *Ann. NY Acad. Sci. 959*: 113–116.

Mattson, M. P., Duan, W., Lee, J., & Guo, Z. (2001a). Suppression of brain aging and neurodegenerative disorders by dietary restriction and environmental enrichment: molecular mechanisms. *Mech. Ageing Dev. 122*: 757–778.

Mattson, M. P., Duan, W., Lee, J., Guo, Z., Roth, G. S., Ingram, D. K., & Lane, M. A. (2001b). Progress in the development of caloric restriction mimetic dietary supplements. *J. Anti-Aging Med. 4*: 225–232.

McCarter, R. J. M., Shimokawa, I., Ikeno, Y., Higami, Y., Hubbard, G. B., Yu, B. P., & McMahan, C. A. (1997). Physical activity as a factor in the action of dietary restriction on aging: Effects in Fischer 344 rats. *Aging Clin. Exp. Res. 9*: 73–79.

McCarty, M. F. (1993), Homologous physiological effect of phenformin and chromium picolate. *Med. Hypothesis 41*: 316–324.

Migliaccio, E., Giorgio, M., Mele, S., Pelicci, G., Reboldi, P., Pandolfi, P. P., Langfrancone, L., & Pelicci, P. G. (1999). The p66[shc] adaptor protein controls oxidative stress and life span in mammals. *Nature 402*: 309–313.

Milgram, N. W., Racine, R. J., Nellis, P., Mendonca, A., & Ivy, G. O. (1990). Maintenance on L-deprenyl prolongs life in aged male rats. *Life Sci. 47*: 415–420.

Nelson, J. F. (1995). The potential role of selected endocrine systems in aging processes. In: E. J. Masoro (Ed.), *Handbook of Physiology*, Section 11, *Aging* (pp. 377–394). New York: Oxford University Press.

Piantanelli, L., Zaia, A., Rossolini, G., Vitichhi, C., Testa, R., Basso, A., & Antognini, A. (1994). Influence of L-deprenyl treatment on mouse survival kinetics. *Ann. NY Acad. Sci. 717*: 72–78.

Pierpaoli, W. & Regelson, W. (1994). Pineal control of aging: effect of melatonin and pineal grafting on aging mice. *Proc. Natl. Acad. Sci. USA 91*: 787–791.

Retzlaff, E., Fontaine, J., & Furata, W. (1966). Effect of daily exercise on lifespan of albino rats. *Geriatrics 21*: 171–177.

Roth, G. S., Ingram, D. K., & Lane, M. A. (2001). Caloric restriction in primates and relevance to humans. *Ann. NY Acad. Sci. 928*: 305–315.

Rowe, J. W. & Kahn, R. L. (1998). *Successful Aging*. New York: Pantheon Books.

Ruehl, W. W., Entriken, T. L., Muggenburg, B. A., Bruyette, D. S., Groffith, W. C., & Hahn, F. F. (1997). Treatment with L-deprenyl prolongs life in elderly dogs. *Life Sci. 61*: 1037–1044.

Saito, K., Yoshioka, H., & Cutler, R. G. (1998). A spin trap, N-tert-butyl-alpha-phenylnitrone extends the life span of mice. *Biosci. Biotechnol. Biochem. 62*: 792–794.

Schwartz, A. G. (1979). Inhibition of spontaneous breast cancer formation in female C3H(Avy/a) mice by long-term treatment with dehydroepiandrosterone. *Cancer Res. 39*: 1129–1132.

Sell, D. R., Nelson, J. F., & Monnier, V. M. (2001). Effect of chronic aminoguanidine treatment on age-related glycation, glycoxidation, and collagen cross-linking in the Fischer 344 rat. *J. Gerontol.: Biol. Sci. 56A*: B405–B411.

Slonaker, J. R. (1912). The normal activity of the albino rat from birth to natural death, its rate of growth, and duration of life. *J. Anim. Behav. 2*: 20–42.

Stoll, S., Hafner, U., Kranzlin, B., & Muller, W. E. (1997). Chronic treatment of Syrian hamsters with low-dose selegiline increases life span in females but not males. *Neurobiol. Aging 18*: 205–211.

Stoll, S., Hartman, H., Cohen, S. A., & Muller, W. E. (1993). The potent free radical scavenger α-lipoic acid improves memory in aged mice: Putative relationship to NMDA receptor deficits. *Pharmacol. Biochem. Behav. 46*: 799–805.

Weindruch, R., Keenan, K. P., Carney, J. M., Fernandes, G., Feuers, R. J., Floyd, R. A., Halter, J. F., Ramsey, J. J., Richardson, A., Roth, G. S., & Spindler, S. R. (2001). Caloric restriction mimetics: Metabolic interventions. *J. Gerontol. 56A*: 20–33.

Weindruch, R., McFeeters, G., & Walford, R. L. (1984). Food intake reduction and immunologic alterations in mice fed dehydroepiandosterone. *Exp. Gerontol. 19*: 297–304.

Yen, T. T. & Knoll, J. (1992). Extension of lifespan in mice treated with DINHLANG (Policias Fruticosum) and (−)deprenyl. *Acta Physiol. Hung. 79*: 119–124.

Yu, B. P. (1995). Putative Interventions. In: E. J. Masoro (Ed.), *Handbook of Physiology, Section 11, Aging* (pp. 613–631). New York: Oxford University Press.

Index